STANDARD TERMS
OF THE ENERGY ECONOMY

Other titles of interest

STANDARD TERMS
OF THE ENERGY ECONOMY

A Glossary for Engineers, Research Workers, Industrialists and
Economists Containing over 600 Standard Energy Terms in
English, French, German and Spanish

The World Energy Conference,
34 St. James's St.,
London SW1A 1HD

PERGAMON PRESS
OXFORD · NEW YORK · TORONTO · SYDNEY · PARIS · FRANKFURT

U.K.	Pergamon Press Ltd., Headington Hill Hall, Oxford OX3 0BW, England
U.S.A.	Pergamon Press Inc., Maxwell House, Fairview Park, Elmsford, New York 10523, U.S.A.
CANADA	Pergamon of Canada Ltd., 75 The East Mall, Toronto, Ontario, Canada
AUSTRALIA	Pergamon Press (Aust.) Pty. Ltd., 19a Boundary Street, Rushcutters Bay, N.S.W. 2011, Australia
FRANCE	Pergamon Press SARL, 24 rue des Ecoles, 75240 Paris, Cedex 05, France
FEDERAL REPUBLIC OF GERMANY	Pergamon Press GmbH, 6242 Kronberg-Taunus, Pferdstrasse 1, Federal Republic of Germany

First edition 1978

British Library Cataloguing in Publication Data

Standard terms of the energy economy.
1. Power resources - Dictionaries - Polyglot 2. Dictionaries, Polyglot
I. Ruttley, E
621.4   TJ163.2   78-40304
ISBN 0-08-022445-8

Printed in Great Britain by Netherwood Dalton & Co. Ltd., Huddersfield

WORLD ENERGY CONFERENCE—OFFICERS

President: Mr. A. G. MUTDOĞAN (Turkey)

International Executive Council

Chairman: Herr Prof. Dres. H. MANDEL (Federal Republic of Germany)

Vice-Chairmen: Mr. G. M. MACNABB (Canada)
Mr. M. T. DIAWARA (Ivory Coast)
Mr. E. GRAFSTRÖM (Sweden)

Secretary-General: Mr. E. RUTTLEY

Members of the Working Group on Standard Terms of the Energy Economy

Co-ordinator: Monsieur M. KLEINPETER (France)

Members: Dr. L. BAUER (Austria)
Dipl.-Ing. E. DUIS (Federal Republic of Germany)
Dipl. Ing. oec. H. OCROB (German Democratic Republic)
Mr. L. H. LEIGHTON (Great Britain)
Mr. A. J. VAN RIEMSDYK (Great Britain)
Ing. S. G. DE VINUESA (Spain)

Contents

Foreword

This first collection contains slightly fewer than 600 terms relating to energy, together with their definitions. The problem of the standardisation of units of measurement having been solved by the introduction of SI units, it seemed similarly desirable to move towards greater harmonisation in energy terminology, especially as it had become apparent that in reports presented at the meetings of the World Energy Conference different terms were being employed to describe identical concepts.

It was on the initiative of the German-speaking countries that the preliminary work was undertaken in 1971.

With the agreement of the International Executive Council, a committee comprising representatives of the French, German, Spanish and English language National Committees has prepared a uniform set of terms and definitions that forms the first part of a reference work designed to appeal to a wide readership. Corresponding terms in Russian, Japanese and Portuguese are being prepared.

The directives laid down by the International Executive Council to serve as guidelines to the Committee entrusted with the preparation of this glossary were as follows:

(1) For those terms where authoritative translations already exist in specialist multi-lingual energy dictionaries, the translations to be found in those dictionaries should be used.
(2) Synonyms should be permitted where appropriate.
(3) Definitions should be produced for terms where the lack of definition may lead to uncertainty of meaning.

The authors of the present glossary are aware that their work is incomplete. The International Executive Council has given authority for the work to be continued in a second stage which will comprise:

(1) The preparation of a section relating to energy conservation.
(2) The preparation of a section on unconventional energy sources (solar energy, geothermal energy, wind energy, etc.).
(3) Additions to various sections of the existing glossary.

In the opinion of the authors the completion of the present edition represents an important first step; they hope that it may prove to be of assistance to the engineer and to others professionally involved with energy in better comprehension through the use of standardised terminology of which there will exist agreed translations in several languages.

Avant - Propos

Le premier recueil de termes contient un peu moins de 600 définitions d'économie énergétique. Après la solution du problème des unités par l'adoption du système S.I., il semblait souhaitable d'aboutir dans le domaine de l'énergie à des concepts plus harmonisés, d'autant qu'on avait constaté que dans les rapports présentés à la Conférence Mondiale de l'Energie, des expressions différentes étaient utilisées dans des contextes identiques.

A l'initiative des pays de langue allemande, un travail préparatoire avait été effectué dès 1971.

Avec l'accord du Conseil exécutif international, un Comité groupant des représentants des Comités nationaux de langues anglaise, française, allemande et espagnole a finalement préparé un ensemble des définitions qui représente la première partie d'un ouvrage appelé à une large diffusion. En outre, la liste des termes en russe, en japonais et en portugais est en préparation.

Les principes directeurs énoncés par le Conseil exécutif international ont servi de guide au Comité d'étude chargé d'établir ce document, à savoir:

(1) Les traductions bien établies de termes mentionnées dans des dictionnaires multilingues relatifs à l'énergie, doivent être utilisées dans le présent ouvrage.
(2) Des synonymes appropriés peuvent être utilisés.
(3) Des définitions doivent être données pour des termes si l'absence de définition peut entraîner des confusions.

Les auteurs du présent vocabulaire sont conscients que leur travail n'est pas terminé. Le Conseil exécutif international a autorisé la poursuite des travaux dans une deuxième phase comprenant:

(1) L'élaboration d'un chapitre relatif à la conservation de l'énergie.
(2) L'élaboration d'un chapitre sur les énergies non conventionnelles (énergie solaire, géothermique, énergie du vent etc.).
(3) Les compléments à apporter aux différentes sections du présent vocabulaire.

Les auteurs pensent qu'un premier pas important a été fait et espèrent mettre à la disposition de l'ingénieur et de l'économiste, un instrument qui leur servira à voir plus clair dans un domaine où règnent encore certaines confusions.

Ce recueil multilingue permettra également de mieux se comprendre entre collègues de pays différents.

Vorwort

Die erste Begriffesammlung enthält fast 600 Definitionen der Energiewirtschaft. Nach der Lösung des Problems der Einheiten durch die Annahme des S.I. Systems schien es wünschenswert, auf dem Gebiet der Energiewirtschaft besser übereinstimmende Begriffe anzuwenden, nachdem man festgestellt hat, daß insbesondere in den, der WEK vorgelegten Berichten verschiedene Ausdrücke für ein und denselben Begriff verwendet wurden.

Auf Initiative der deutschsprachigen Länder wurden seit 1971 hiezu vorbereitende Arbeiten geleistet.

Mit der Zustimmung des Internationalen Exekutivrates hat ein Ausschuß, der sich aus Vertretern der englischen, französischen, deutschen und spanischen Nationalkomitees zusammensetzt, schließlich eine Zusammenfassung von Definitionen, die den 1. Teil einer, für eine weite Verbreitung bestimmten, Arbeit repräsentiert, vorbereitet. Entsprechende Ausgaben in Russisch, Japanisch und Portugisisch sind in Vorbereitung.

Die vom Internationalen Exekutivrat festgelegten Grundsätze dienten dem Ausschuß als Richtlinien, wobei folgende Punkte zu berücksichtigen waren:

(1) Bei denjeniegen Ausdrücken für die in mehrsprachigen Wörterbücher des Energiebereiches Übersetzungen schon bestehen, sind letztere im vorliegenden Werk zu benutzen.
(2) Passende Synonyme sind erlaubt.
(3) Kann ein Fehlen einer Begriffsbestimmung zu Verständnisschwierigkeiten führen, soll diese Begriffsbestimmung ausgearbeitet werden.

Die Verfasser der vorliegenden Zusammenstellung sind sich bewußt, daß ihre Arbeit nicht beendet ist. Der Internationale Exekutivrat hat die Fortsetzung der Arbeiten in einer zweiten Phase angeordnet. Diese beinhaltet:

(1) Ausarbeitung eines Begriffsfeldes bezüglich des Haushaltens mit Energie.
(2) Ausarbeitung eines Begriffsfeldes über die Nichtkonventionellen Energiequellen (Sonnenenergie, geothermische Energie, Windenergie, etc.).
(3) Die verschiedenen Begriffsfelder der vorliegenden Zusammenstellung zu ergänzen.

Die Verfasser glauben, daß mit dieser ersten Ausgabe ein wichtiger Schritt gemacht wurde. Sie hoffen, dem Ingenieur und Volkswirt damit ein Instrument zur Verfügung zu stellen zum besseren Verständnis im Gebrauch einer einheitlichen Terminologie, für welche bereits vereinbarte Übersetzungen in einigen Sprachen bestehen.

Prefacio

Esta primera recopilación de términos contiene algo menos de 600 definiciones relativas a la economía energética. Una vez resuelto el problema de las unidades, al haber sido adoptado el sistema S.I., parecía deseable llegar, en el dominio de la energía, a conceptos más armonizados, tanto más cuanto que se ha comprobado que, en los informes presentados a la Conferencia Mundial de la Energía, vienen siendo utilizadas expresiones distintas en contextos idénticos.

Siguiendo la iniciativa de los países de habla alemana, se viene realizando un trabajo preparatorio a partir del año 1971.

Contando con la conformidad del Consejo Ejecutivo Internacional, un Comité, que agrupa representantes de los comités nacionales de lengua inglesa, francesa, alemana y española, ha preparado un conjunto unificado de términos y definiciones que representa la primera parte de un trabajo destinado a alcanzar una gran difusión. Se encuentran en preparación términos correspondientes en ruso, en japonés y en portugués.

El Comité de estudios encargado de redactar este documento ha utilizado, como guía, los siguientes principios directores, que fueron enunciados por el Consejo Ejecutivo Internacional:

1° Aquellos términos, para los que ya existían traducciones con garantía en diccionarios multilingües especializados en términos energéticos, son los que deben ser empleados.
2° Los sinónimos deben ser permitidos cuando resulten apropiados.
3° Deben ser definidos aquellos términos en que la falta de definición puede originar una incertidumbre en su significado.

Los autores de este vocabulario reconocen que su trabajo no se ha terminado. El Consejo Ejecutivo Internacional ha autorizado la continuación de los trabajos en una segunda fase que comprende:

1° La elaboración de un capítulo relativo a la conservación de la energía.
2° La elaboración de un capítulo sobre energías no convencionales (energía solar, geotérmica, eólica etc.).
3° Los complementos que hayan de ser aportados a las diferentes secciones del presente vocabulario.

Los autores estiman que la terminacíon de esta edición representa un primer paso importante y tienen la esperanza de poner a disposición del ingeniero y del economista un instrumento que, empleando una terminologia uniforme con traducciones convenidas en varios idiomas, permitirá una mejor comprensión.

Acknowledgements

During the chairmanship of my predecessor, Monsieur Roger Gaspard, the proposal of the German-speaking National Committees to prepare a multi-lingual vocabulary of standardised terms and definitions of the energy economy was accepted by the International Executive Council. Representatives of seven countries—Austria, France, the Federal Republic of Germany, the German Democratic Republic, Spain, Great Britain and Switzerland—joined together in collaboration to produce a list of terms and definitions and their equivalents in English, French, German and Spanish. While Great Britain acted as a liaison with the English-speaking countries of the world and Spain organised the collaboration of and contribution from Spanish-speaking countries, it was equally natural that Switzerland could act as a bridge between German and French language speaking countries. The Swiss National Committee also at our wish appointed a Coordinator for the operation. We owe a great debt to the late Monsieur Albert Ebener of Switzerland who from 1975 onwards brought together and harmonised the efforts of these various linguistic groups. To our great sorrow M. Ebener died in the summer of 1977 when the final completed version in four languages was ready to be approved by the International Executive Council in Istanbul. The present volume will remain as a testimony to his dedication, tact and persuasion in bringing into ordered form the work of many countries.

The preparation of this work has stimulated other countries, notably Japan, the USSR, Portugal, Iceland and the Arab-speaking world, to join in this co-operative effort and to prepare equivalent terms in their own languages. Meanwhile a reconstituted working group is extending the work of this first volume to include other sectors of the energy field.

In this undertaking we have benefited from the support and counsel of the International Gas Union and the Union International des Producteurs et Distributeurs d'Energie Electrique and to their respective presidents and members I acknowledge our gratitude for their willingness to put material at our disposal in the preparation of this volume. I also acknowledge with gratitude the efforts of all those from many countries who have given generously of their time and effort in bringing together and refining the lists of terms and definitions which make up this first volume.

I believe that in collaborative efforts of this kind designed to persuade energy practitioners to employ standardised terminology readily understood internationally, the World Energy Conference is performing a valuable task in thus promoting better communication amongst nations in the energy sphere.

Remerciements

C'est sous la présidence de mon prédécesseur, Monsieur Roger Gaspard, que le Conseil Exécutif International a approuvé l'idée lancée par les Comités nationaux germanophones de préparer un lexique multilingue des termes de l'économie énergétique ainsi que de leurs définitions normalisées. Les représentants de sept pays: Autriche, France, République Fédérale d'Allemagne, République Démocratique Allemande, Espagne, Grande Bretagne et Suisse, ont collaboré afin de produire une liste de termes et de définitions en anglais, en français et en espagnol. Tandis que la Grande Bretagne assurait la liaison avec les autres pays anglophones du monde et l'Espagne a organisé la collaboration et la contribution des pays de langue espagnole, il était tout aussi normal que la Suisse coordonne les travaux entre les pays francophones et germanophones. A notre demande, le comité national suisse a également nommé un coordonateur chargé de mener cette opération. Nous avons une grande dette de reconnaissance envers feu Monsieur Albert Ebener, du Comité national Suisse qui, à partit de 1975, a coordonné et harmonisé les travaux des différents groupes linguistiques. A notre grand regret, Monsieur Ebener est décédé dans le courant de l'été 1977, au moment même où s'achevait la version définitive en quatre langues qui devait être peu après soumise pour approbation au Conseil Exécutif International à Istanbul. Le présent volume restera un témoignage des efforts qu'il y a consacrés, du tact et du don de persuasion dont il a fait preuve pour mener à bien la coordination des travaux de tant de pays.

La préparation de ce volume a stimulé d'autres pays, notamment le Japon, l'URSS, le Portugal, l'Islande et les pays arabophones, qui ont décidé de se joindre à cette coopération internationale et d'élaborer des ouvrages équivalents dans leurs langues respectives. Entre temps, un groupe de travail a été reconstitué pour compléter le premier volume par l'inclusion d'autres secteurs du domaine de l'ènergie.

Dans cette entreprise, nous avons profité du soutien et des conseils de l'Union Internationale du Gaz et de l'Union Internationale des Producteurs et Distributeurs d'Energie Electrique. Je tiens à exprimer aux Présidents et aux membres de ces Organisations toute notre reconnaissance pour l'aide et les informations qu'ils ont bien voulu nous fournir au cours de la préparation de cet ouvrage. Je tiens également à remercier tous ceux qui, dans de nombreux pays, ont consacré généreusement leur temps et leurs efforts à la compilation des termes et des définitions et ont ainsi permis la réalisation de ce premier volume.

Je suis convaincu qu'en s'attachant à réaliser une coopération internationale de ce genre, dont le but est la normalisation internationale de la terminologie énergétique, la Conférence Mondiale de l'Energie accomplit une tâche utile et contribue à la promotion d'une meilleure communication entre tous les pays dans le domaine de l'énergie.

Anerkennung

Während der Amtszeit meines Vorgängers, Herrn Roger Gaspard, gab der Internationale Exekutivrat seine Zustimmung zu dem Plan der deutschsprachigen Nationalen Komitees, eine mehrsprachige Terminologie standardisierter Begriffe und Definitionen der Energiewirtschaft zusammenzustellen. Vertreter von sieben Ländern—der Bundesrepublik Deutschland, der DDR, Frankreichs, Großbritanniens, Österreichs, der Schweiz und Spaniens—setzten sich zusammen und erarbeiteten gemeinsam eine Liste von Begriffen und Definitionen mit den entsprechenden Fachbezeichnungen in deutscher, englischer, französischer und spanischer Sprache. Da Großbritannien als Vermittler zwischen den englisch sprechenden Ländern der Welt fungierte und Spanien organisierte Federführend die Mitarbeit und zweitens die Beitrage der spanisch-französisch sprechenden Ländern übernahm. Auf unseren Wunsch ernannte das Schweizer Nationale Komitee außerdem einen Koordinator für dieses Projekt. Wir sind dem verstorbenen Herrn Albert Ebener aus der Schweiz zu großem Dank verpflichtet. Von 1975 an hat er die Arbeiten der verschiedenen Sprachengruppen koordiniert und in Einklang gebracht. Leider verstarb Herr Ebener im Sommer 1977, als das Werk in seiner endgültigen Fassung in vier Sprachen zur Annahme durch den Internationalen Exekutivrat in Istanbul vorlag. Dieser Band, in dem er die Arbeit vieler Länder in eine geordnete Form brachte, ist ein Beweis für seine Einsatzbereitschaft, sein Einfühlungsvermögen und seine Überzeugungskraft.

Die Ausarbeitung dieses Werkes veranlaßte auch andere Staaten, vornehmlich Japan, die UdSSR, Portugal und die arabisch sprechenden Länder, diesem Beispiel der Zusammenarbeit zu folgen und in ihren eigenen Sprachen entsprechende Begriffsbestimmungen zusammenzustellen. Inzwischen arbeitet eine Arbeitsgruppe an der Erweiterung dieses ersten Bandes, um auch andere Bereiche der Energiewirtschaft einzuschließen.

Bei dieser Arbeit erhielten wir die freundliche Unterstützung und den Rat der Internationalen Gas-Union und der Union Internationale des Producteurs et Distributeurs d'Energie Electrique. Unser Dank und unsere Anerkennung gilt den Präsidenten und Mitgliedern dieser beiden Verbände, die uns für die Erarbeitung dieses Buches so bereitwillig Unterlagen zur Verfügung gestellt haben. Ebenfalls zu großem Dank verbunden bin ich all denen in der Welt, die durch tatkräftige Mitarbeit bei der Zusammenstellung und Überarbeitung der Listen von Begriffen und Definitionen dieses ersten Bandes weder Zeit noch Mühe gescheut haben.

Diese Zusammenarbeit hat ein Ziel vor Augen: die Fachleute aus der Energiewirtschaft zur Anwendung einer standardisierten und international leicht verständlichen Terminologie zu bewegen. Ich glaube daß die Weltenergiekonferenz mit dieser Arbeit einen wertvollen Beitrag zur besseren Verständigung der Völker im Energiebereich leistet.

Reconocimiento

Bajo la presidencia de mi predecesor, Monsieur Roger Gaspard, el Consejo Ejecutivo Internacional aceptó la proposición de los Comités Ejecutivos de habla alemana, de preparar un vocabulario plurilingüe de conceptos relativos a la economía energética y sus definiciones. Representantes de siete países Austria, Francia, República Federal Alemana, República Democrática Alemana, España, Gran Bretaña y Suiza, han colaborado conjuntamente en la redacción de una lista de términos y definiciones y sus equivalentes en inglés, francés, alemán y español. Mientras Gran Bretaña actuaba coordinando los países de habla inglesa y España contaba con la colaboración y delegación de los hispanoparlantes, era natural que Suiza actuase como puente entre los países de habla alemana y los francófonos. A petición nuestra también el Comité Nacional Suizo designó un coordinador a estos efectos. Hemos contraído una gran deuda con el fallecido M. Albert Ebener de Suiza, quien, a partir de 1975 reunió y armonizó los esfuerzos de los distintos Grupos lingüísticos.

Con gran sentimiento nuestro, Mr. Ebener falleció en el verano de 1977, cuando la versión final completa en cuatro idiomas estaba preparada para su aprobación por el Consejo Ejecutivo Internacionalen Estambul. Este volúmen será un testimonio de su dedicación, tacto y capacidad de persuasión para conseguir poner en órden los trabajos de numerosos países.

La preparación de este trabajo ha servido de estímulo a otros pueblos, principalmente Japón, la URSS, Portugal, Islandia y el mundo Arabe, para unirse a este esfuerzo cooperativo y preparar los términos equivalentes en sus propios idiomas. Mientras tanto un grupo de trabajo reconstituído, procede a ampliar la labor de este primer volumen con el fin de incluir otros sectores del campo de la energía.

En este empeño nos hemos beneficiado de la ayuda y consejo de la Unión Internacional del Gas y de la Unión Internacional de Productores y Distribuidores de Energía Eléctrica. Hago llegar nuestro agradecimiento a sus respectivos presidentes y miembros por la buena voluntad demostrada poniendo a nuestra disposición su material informativo para preparar este volúmen. Asimismo reconozco con gratitud, los esfuerzos de todas aquellas personas de muchos países que han hecho entrega generosa de su tiempo y esfuerzo, resumiendo y perfeccionando las relaciones de términos y definiciones reunidos en este primer volúmen.

Creo que la Conferencia Mundial de la Energia está realizando una importante tarea al promover una mejor comunicación entre las naciones en la esfera de la energía, gracias a los esfuerzos de esta clase de colaboración, destinadas a persuadir al empleo de una terminología normalizada y que se entienda internacionalmente con facilidad, a quienes actuan en el campo energético.

Section 1 **1**

General Terms
Concepts Généraux
Allgemeine Begriffe
Conceptos Generales

General Terms

1.1 General Terms

1.1.1 The energy industries; the energy sector; the energy economy That part of the national economy that is concerned in meeting a nation's energy requirements.

1.1.2 Natural energy The total amount of the energy occurring in nature and recoverable by technical means.

1.1.3 Forms of energy; sources of energy The meaning of these terms is clear from the terms falling under this heading. In English usage the term "form of energy" generally refers to the physical form in which energy occurs (heat, kinetic energy, electricity), whereas "source of energy" generally refers to the potential to release energy (coal, oil).

1. **Water power; hydro-electric power** (see Section 3)
2. **Solid fuels** (see Section 4)
3. **Liquid fuels** (see Section 5)
4. **Gaseous fuels** (see Section 6)
5. **Nuclear energy** (see Section 7)
6. **Solar energy** That part of the energy from solar radiation that can be usefully recovered.
7. **Wind energy** The energy that can be usefully recovered from the wind. More generally termed "wind power".
8. **Tidal energy** The energy that can be usefully recovered by exploiting the difference in water levels due to the ebb and flow of the tides. More generally termed "tidal power".
9. **Geothermal energy** The thermal energy transferred to water or steam from molten underground rock.
10. **Waste fuels** Combustible waste materials whose calorific value is utilised for energy production.
11. **Waste heat** By-product heat energy that occurs unavoidably in industrial processes.

1.1.4 Primary energy; crude energy Energy that has not been subjected to any conversion or transformation process.

1.1.5 Secondary energy; derived energy Energy that has been produced by the conversion or transformation of primary energy or of another secondary form of energy.

1.1.6 Energy supplied; energy available The energy made available to the consumer before its final conversion (i.e. before final utilisation).

1.1.7 Useful energy; net energy The energy made usefully available to the consumer in its final conversion (i.e. in its final utilisation).

1.1.8 Sources of energy All sources from which useful energy can be recovered directly or by means of a conversion or transformation process. The terms "sources of energy", "forms of energy" and "energy" are interchangeable in many contexts. See under 1.1.3 above.

1.1.9 Energy transformation The recovery or production of energy involving a physical change of state of the form of energy (e.g. coal liquefaction). In English usage the term "energy conversion" is commony employed in both this sense and in the sense given in 1.1.10 below.

Concepts Généraux

1.1 Termes Généraux

1.1.1 Economie énergétique Partie de l'économie qui traite de la couverture des besoins en énergie.

1.1.2 Energie naturelle (potentiel énergétique) Totalité des énergies existant dans la nature et exploitables par des moyens techniques.

1.1.3 Formes d'énergie primaire

1. **Energie hydraulique** (voir Section n° 3)
2. **Combustibles solides** (voir Section n° 4)
3. **Combustibles liquides** (voir Section n° 5)
4. **Combustibles gazeux** (voir Section n°6)
5. **Energie nucléaire** (voir Section n° 7)
6. **Energie solaire** Part de chaleur exploitable du rayonnement solaire.
7. **Energie éolienne** Energie exploitable par l'utilisation des vents.
8. **Energie marémotrice** Energie exploitable par l'utilisation des variations du niveau de la mer dues aux marées.
9. **Energie géothermique** Energie thermique contenue dans l'eau ou la vapeur d'eau et provenant de roches souterraines en fusion.
10. **Combustibles résiduaires** (résidus industriels) Déchets combustibles dont le potentiel calorifique est utilisé pour la production d'énergie.
11. **Energie résiduelle** Energie que les processus techniques produisent inévitablement sous forme de chaleur perdue.

1.1.4 Energie primaire (énergie brute) Energie n'ayant subi aucune conversion.

1.1.5 Energie secondaire Energie provenant de la conversion d'énergie primaire ou d'autres énergies secondaires.

1.1.6 Energie disponible Energie fournie au consommateur avant la dernière conversion.

1.1.7 Energie utile Energie dont dispose le consommateur après la dernière conversion.

1.1.8 Sources d'énergie Tout ce qui permet de produire de l'énergie utile directement ou par conversion ou transformation.
Du point du vue de l'économie énergétique, l'énergie et les sources d'énergie sont des concepts synonymes.

1.1.9 Conversion d'énergie Production d'énergie avec modification de l'état physique de l'agent énergétique.

Allgemeine Begriffe

1.1 Allgemeine Begriffe

1.1.1 Die Energiewirtschaft Der Teil der Wirtschaft, welcher sich mit der Deckung des Energiebedarfes befasst.

1.1.2 Naturenergie (Energiepotential) Die Gesamtheit der in der Natur vorhandenen und mit technischen Mitteln gewinnbaren Energien.

1.1.3 Primäre Energieformen
.1 **Wasserkraft** (siehe Begriffsfeld 3)
.2 **Feste Brennstoffe** (siehe Begriffsfeld 4)
.3 **Flüssige Brennstoffe** (siehe Begriffsfeld 5)
.4 **Gasförmige Brennstoffe** (siehe Begriffsfeld 6)
.5 **Kernenergie** (siehe Begriffsfeld 7)
.6 **Sonnenergie** ist der gewinnbarer Wärmeanteil der Sonnenstrahlung.
.7 **Windenergie** Die gewinnbare Energie von Luftbewegungen relativ zur Erdoberfläche.
.8 **Gezeitenenergie** Die Energie, die sich bei Ausnützung der durch Ebbe und Flut entstehenden Schwankungen des Meeresspiegels gewinnen lasst.
.9 **Erdwärme** Die an Wasser oder Wasserdampf gebundene, aus unterirdischen feuerflüssigen Gesteinen stammende Wärmeenergie.
.10 **Abfallbrennstoffe** sind brennbare Abfälle, deren Heizwert energiewirtschaftlich genutzt wird.
.11 **Abfallenergie** Die bei technischen Prozessen zwangsläufig als Abwärme anfallende Energie.

1.1.4 Primärenergie (Rohenergie) Energie, die keiner Umwandlung unterworfen wurde.

1.1.5 Sekundärenergie Energie, die aus der Umwandlung von Primärenergie oder aus anderer Sekundärenergie gewonnen wurde.

1.1.6 Gebrauchsenergie Die Energie, welche dem Verbraucher vor der letzten Umwandlung zur Verfügung gestellt wird.

1.1.7 Nutzenergie Die Energie, welche beim Verbraucher nach der letzten Umwandlung zur Verfügung steht.

1.1.8 Energieträger Alles, woraus Nutzenergie direkt oder durch Umformung oder Umwandlung gewonnen werden kann.
Aus der Sicht der Energiewirtschaft sind Energie und Energieträger synonyme Begriffe.

1.1.9 Energieumwandlung Die Gewinnung von Energie unter Aenderung der physikalischen Erscheinungsform des Energieträgers.

Conceptos Generales

1.1. Conceptos Generales

1.1.1 Economía energética Parte de la economía relativa a las necesidades de energía.

1.1.2 Energía natural (potencial energético) Cantidad total de energía presente en la naturaleza, que se puede obtener por medios técnicos.

1.1.3 Formas de energía primaria
1. **Energía hidráulica** (Ver Sección 3).
2. **Combustibles sólidos** (Ver Sección 4).
3. **Combustibles líquidos** (Ver Sección 5).
4. **Combustibles gaseosos** (Ver Sección 6).
5. **Energía nuclear** (Ver Sección 7).
6. **Energía solar** Es la parte aprovechable de la radiación solar.
7. **Energía eólica** Energía aprovechable por utilización de los vientos.
8. **Energía maremotriz** Energía aprovechable utilizando las variaciones, debidas a las mareas, del nivel del mar.
9. **Energía geotérmica** Energía térmica contenida en el agua o en el vapor de agua, procedente de las rocas subterráneas en fusión.
10. **Combustibles de desechos** (residuos industriales) son desechos combustibles cuyo potencial calorífico se utiliza para producir energía.
11. **Energía calorífica residual** Es un subproducto de la energía calorífica que se produce inevitablemente en los procesos industriales.

1.1.4 Energía primaria (energía bruta) Energía que no ha sido sometida a ningún proceso de conversión.

1.1.5 Energía secundaria Energía procedente de la conversión de energía primaria o de otras energías secundarias.

1.1.6 Energía disponible Energía suministrada al consumidor antes de su conversión final.

1.1.7 Energía util Energía de que dispone el consumidor después de su última conversión.

1.1.8 Fuente de energía Todo aquello que permite producir energía útil directamente o por medio de conversión o transformación. Desde el punto de vista de la economía energética los términos "fuente de energía" y "energía" son sinónimos.

1.1.9 Conversion de energía Producción de energía con modificación del estado físico del agente energético.

1

1.1.10 Energy conversion The recovery or production of energy involving no change in the physical state of the form of energy (e.g. coke from coal).

1.1.11 Energy utilisation Obtaining useful energy from the energy supplied.

1.2 Energy Balance Terms

1.2.1 Energy balance A quantitative statement referred to a specific economic area, system or process for a specified period of time, of the energy input on the one side and energy consumption on the other, the statement including losses occurring in conversion and transport as well as input of forms of energy that are not utilised for energy purposes. The term "heat balance" is analogous.

1.3 Terms Relating to Time and Capacity

1.3.1 Operating time The period of time during which a plant or part of a plant supplies useful energy.

1.3.2 Stand-by availability time; stand-by time; reserve shutdown availability time; reserve shutdown time The period of time during which a plant or a part of a plant could supply useful energy after the normal period of start-up.

1.3.3 Planned unavailability time; planned outage time; planned down time The period of time during which a plant or part of a plant is not in running order due to planned maintenance.

1.3.4 Unplanned unavailability time; unplanned outage time; unplanned down time The period of time during which a plant or part of a plant is not in running order due to unforeseen breakdown.

1.3.5 Availability time The sum of the operating time and the stand-by availability time, etc.

1.3.6 Unavailability time; outage time; down time The total of the planned and unplanned unavailability time, etc.

1.3.7 Reference period The period of time to which data relate; in the context of this Section it is the sum of the availability time and the unavailability time, etc.

1.3.8 Utilisation period (of maximum demand) The quotient of the energy obtained, produced, distributed or consumed within a specific period and the maximum capacity of (or demand on) the plant occurring within the same period.

1.3.9 Availability time ratio When referred to a plant or part of a plant, the ratio of the availability time to the reference period.

1.3.10 Operating time ratio The ratio of the operating time to the reference period.

1.3.11 Nominal capacity; rated capacity; rated power; rating The maximum continuous capacity/power/rating for which the plant has been ordered and designed, as indicated on the makers' nameplate or in the manufacturers' specification.

1.3.12 Nominal generation; nominal production The product of the nominal capacity and the reference period. The term "nominal output" is sometimes employed; the word "output", however, is imprecise and may mean either production or capacity.

1.1.10 Transformation d'énergie Production d'énergie avec conservation de l'état physique de l'agent énergétique.

1.1.11 Mise en oeuvre de l'énergie Production d'énergie utile à partir d'énergie disponible.

1.2 Concepts des termes des bilans

1.2.1 Bilan énergétique Relevé statistique des ressources et de l'utilisation de sources d'énergie à l'intérieur d'une zone économique déterminée pour une période déterminée, compte tenu des pertes résultant de la conversion, de la transformation et du transport ainsi que des ressources d'énergie, servant à des fins sans rapport avec l'économie de l'énergie.

1.3 Concepts des Temps et des Puissances

1.3.1 Temps de fonctionnement Durée pendant laquelle une installation ou une partie d'installation fournit de l'énergie utile.

1.3.2 Temps de disponibilité passive Durée pendant laquelle une installation ou une partie d'installation pourrait fournir de l'énergie utile après le temps normal de mise en marche.

1.3.3 Temps d'indisponibilité sur programme (part planifiée du temps d'indisponibilité) Laps de temps durant lequel une installation ou une partie d'installation n'est pas en ordre de marche à la suite d'opérations planifiées.

1.3.4 Temps d'indisponibilité sur avarie (part non planifiée du temps d'indisponibilité) Laps de temps durant lequel une installation ou une partie d'installation n'est pas en ordre de marche à la suite d'un dommage imprévu.

1.3.5 Temps de disponibilité Somme du temps de fonctionnement et du temps de disponibilité passive.

1.3.6 Temps d'indisponibilité Somme du temps d'indisponibiité sur programme et du temps d'indisponibilité sur avarie.

1.3.7 Période de référence Laps de temps auquel se réfère l'indication d'une grandeur.
Elle est équivalente à la somme des temps de disponibilité et d'indisponibilité.

1.3.8 Durée d'utilisation Quotient de l'énergie obtenue, produite., distribuée ou consommée au cours d'une période de temps déterminée par la puissance maximale atteinte au cours de cette même période de temps par l'installation considérée (équipement, appareil).

1.3.9 Facteur de disponibilité d'une installation ou d'une partie d'installation Quotient du temps de disponibilité par la durée de la période de référence.

1.3.10 Facteur d'utilisation Quotient du temps de fonctionnement par la durée de la période de référence.

1.3.11 Puissance nominale Puissance maximale en régime continu pour laquelle l'installation est prévue et dimensionnée.
La puissance nominale doit être recherchée dans le procès-verbal des essais, sur la plaque signalétique ou dans le cahier des charges de construction.

1.3.12 Energie nominale Produit de la puissance nominale par la durée de la période de référence.

1.1.10 Energieumformung Die Gewinnung von Energie unter Wahrung der physikalischen Erscheinungsform des Energieträgers.

1.1.11 Energieanwendung Die Gewinnung von Nutzenergie aus Gebrauchsenergie.

1.2 Bilanzbegriffe

1.2.1 Energiebilanz Der statistiche Nachweis von Aufkommen und Verwendung von Energieträgern innerhalb eines bestimmten Wirtschaftsraumes für eine bestimmte Zeitspanne unter Berücksichtigung der beim Umwandeln, Umformen und Fortleiten auftretenden Verluste sowie des Aufkommens von Energieträgern, die nicht energiewirtschaftlichen Zwecken dienen.

1.3 Zeit- und Leistungsbegriffe

1.3.1 Betriebszeit Die Zeitspanne, während der eine Anlage oder ein Anlageteil nutzbare Energie abgibt.

1.3.2 Bereitschaftszeit Die Zeitspanne, während der eine Anlage oder ein Anlageteil spätestens nach der normalen Anfahrzeit nutzbare Energie abgeben kann.

1.3.3 Reparaturzeit (Plananteil der Nichtverfügbarkeitszeit) Die Zeitspanne, während der eine Anlage oder ein Anlageteil infolge geplanter Eingriffe nicht betriebsbereit ist.

1.3.4 Ausfallzeit (Störanteil der Nichtverfügbarkeitszeit) Die Zeitspanne, während der eine Anlage oder ein Anlageteil infolge eines unvorhergesehenen Schadens nicht betriebsbereit ist.

1.3.5 Verfügbarkeitszeit Die Summe aus Betriebszeit und Bereitschaftszeit.

1.3.6 Nichtvertügbarkeitszeit Die Summe aus Reparaturzeit und Ausfallzeit.

1.3.7 Nennzeit Die Zeitspanne, auf die sich die Angabe einer Grösse bezieht.
Sie ist gleich der Summe aus Verfügbarkeits- und Nichtverfügbarkeitszeit.

1.3.8 Benützungsdauer Gleich dem Quotient aus der in einer bestimmten Zeitspanne gewonnenen, erzeugten, verteilten oder verbrauchten Energie und der von der betrachteten Anlage (Einrichtung, Gerät) erreichten Höchstleistung in der gleichen Zeitspanne.

1.3.9 Zeitverfügbarkeit Einer Anlage oder eines Anlageteils ist gleich dem Quotienten aus der Verfügbarkeitszeit (Betriebszeit plus Bereitschaftszeit) und der Nennzeit.

1.3.10 Zeitausnützung Gleich dem Quotient aus der Betriebszeit und der Nennzeit.

1.3.11 Nennleistung Die höchste Dauerleistung (ohne zeitliche Einschränkung) von Anlagen, für die sie bestellt und bemessen sind. Sie ist auf dem Kennschild angegeben oder aus den Spezifikationen oder aus den Abnahmeprotokollen ersichtlich.

1.3.12 Nennarbeit Das Produkt aus Nennleistung und Nennzeit.

1.1.10 Transformacion de energía Producción de energía, conservando el estado físico del agente energético.

1.1.11 Utilizacion de energía Obtención de energía útil a partir de la energía disponible.

1.2 Conceptos Relativos al Balance Energético

1.2.1 Balance energético Informe estadístico relativo a los recursos de energía dentro de un área económica determinada durante un período determinado de tiempo, teniendo en cuente las pérdidas debidas a la conversión, la transformación y el transporte, así como a los recursos energéticos que sirven a fines sin relación con la economía de la energía.

1.3 Conceptos Relativos al Tiempo y la Potencia

1.3.1 Tiempo de servicio Período de tiempo durante el cual una instalación o parte de ella suministra energía útil.

1.3.2 Tiempo de disponibilidad pasiva Período de tiempo durante el que una instalación, o parte de la misma, podría suministrar energía útil, después del tiempo normal de arranque.

1.3.3 Tiempo de parada programada (Período planificado - del tiempo de parada) Período de tiempo durante el cual una instalación, o una parte de ella no se encuentra en orden de marcha, debido a operaciones planificadas.

1.3.4 Tiempo de parada no programada (Período no planificado, del tiempo de parada) Período de tiempo durante el cual una instalción, o parte de una instalación ne se encuentra en orden de marcha, debido a una avería imprevista.

1.3.5 Tiempo de disponibilidad Es la suma del tiempo de funcionamiento y del tiempo de disponibilidad pasiva.

1.3.6 Tiempo de parada Suma de los tiempos de parada programada y no programada.

1.3.7 Período de referencia Período de tiempo al que se refiere una magnitud determinada - Equivale a la suma de los tiempos de disponibilidad y de parada.

1.3.8 Duración de utilización Cociente entre la energía, obtenida, producida, distribuída o consumida a lo largo de un período de tiempo determinado y la potencia máxima alcanzada, durante este mismo período de tiempo, por la instalación considerada (equipo, dispositivo, máquina. . . .).

1.3.9 Factor de disponibilidad De una instalción o parte de una instalación. Relación del tiempo de disponibilidad al período de referencia.

1.3.10 Factor de utilización La relación entre el tiempo de servicio y el período de referencia.

1.3.11 Potencia nominal Potencia máxima, en régimen continuo, para la que ha sido prevista y dimensionada la instalación. La potencia nominal debe encontrarse en el proceso verbal de los ensayos, en la placa de características o en el pliego de condiciones de construcción de la maquinaria.

1.3.12 Produccion nominal Producto de la potencia nominal por el período de referencia.

1.4 Supply Characteristics

1.4.1 Energy consumption The utilisation of energy for conversion to secondary energy or for the production of useful energy. It should be stated whether the energy consumed is primary energy, secondary energy, energy supplied or useful energy.

1.4.2 Customer The party who receives the energy supplied from the supply or distribution undertaking. In most contexts a wholesale purchaser.

1.4.3 Consumer The party who uses the final energy supplied for his own needs.

1.4.4 Per capita consumption The quotient of the energy consumption of an area and the population of that area.

1.4.5 Security of supply The assurance that energy will be available in the quantities and qualities required under given economic conditions.

1.4 Grandeurs Caractéristiques de l'approvisionnement

1.4.1 Consommation d'énergie Utilisation d'énergie en vue de la conversion en énergie secondaire ou de la production d'énergie utile.
Les niveaux de référence respectifs (énergie primaire, énergie secondaire, énergie disponible, énergie utile) doivent être indiqués.

1.4.2 Clients Personnes physiques ou morales auxquelles est fournie de l'énergie disponible.

1.4.3 Consommateurs d'énergie Personnes physiques ou morales qui utilisent de l'énergie pour leurs propres besoins.

1.4.4 Consommation par habitant Quotient de la consommation d'énergie d'une région par la population résidente de celle-ci.

1.4.5 Sécurité d'approvisionnement Probabilité de disposer à tout moment d'énergie en quantité et en qualité voulues, sous des conditions économiques données.

1.4 Kenngrössen der Versorgung

1.4.1 Energieverbrauch Der Energieeinsatz zur Umwandlung in Sekundärenergie oder zur Gewinnung von Nutzenergie.

Die jeweilige Bezugsebene (Primärenergie, Sekundärenergie, Gebrauchsenergie, Nutzenergie) ist anzugeben.

1.4.2 Abnehmer sind natürliche oder juristische Personen, welche Gebrauchsenergie beziehen.

1.4.3 Verbraucher sind natürliche oder juristische Personen, welche Gebrauchsenergie verwenden.

1.4.4 Pro-Kopf-Verbrauch Gleich dem Quotient aus dem Energieverbrauch eines Gebietes und der Wohnbevölkerung dieses Gebietes.

1.4.5 Versorgungssicherheit Ein Mass für die Wahrscheinlichkeit, dass unter gegebenen wirtschaftlichen Bedingungen jederzeit Energie in der gewünschten Menge und Güte zur Verfügung steht.

1.4 Características de los Suministros

1.4.1 Consumo de energía Utilización de la energía para su conversión en energía secundaria o para la producción de energia útil.

Deben indicarse los niveles de referencia respectivos, es decir, si la energía consumida es energía primaria, energía secundaria, energía final o energía útil.

1.4.2 Cliente, abonado Persona física o jurídica a la que se suministra la energía disponible.

1.4.3 Consumidor Persona o entidad que utiliza la energía para sus propias necesidades.

1.4.4 Consumo por habitante Relación del consumo de energía en una región al número de habitantes que residen en ella.

1.4.5 Seguridad del suministro Posibilidad de disponer en cualquier momento de energía en la cantidad y calidad deseados en determinadas condiciones económicas.

1

Section 2

Electricity Industry
Industrie électrique
Elektrizitätswirtschaft
Industria Electrica

2

Electricity Industry

2.1 Electricity Generating Plant

2.1.1 **Fossil-fuel(l)ed power station** A power station in which the chemical energy contained in solid, liquid and gaseous fuels of fossil origin is converted into electrical energy.

2.1.2 **Nuclear power station** A power station in which the energy released by nuclear fuels is converted into electrical energy. See also 7.1.1.

2.1.3 **Hydro-electric power station** A plant designed to convert the gravitational energy of waters into electrical energy.

2.1.4 **Base-load power station** A power station serving mainly to meet the base load.

2.1.5 **Peak-load power station** A power station serving mainly to meet the peak load.

2.1.6 **Once through water-cooling** A cooling system in which water is drawn from an available source, e.g. river, sea, lake, canal, passed once through the power station condensers and returned in its heated condition directly to the source.

2.1.7 **Cooling with wet cooling towers** A cooling system in which water passing through the power station condensers takes up heat, releases this heat subsequently to atmosphere in a wet cooling tower mainly by evaporation and is then recycled.

2.1.8 **Cooling with dry cooling towers** A cooling system in which heat from the condensers of a power station is dissipated to the atmosphere in a cooling tower solely by convection.

2.1.9 **Power station internal consumption; station service consumption** The electricity consumed by a power station or power station set in its auxiliary plant, including electricity consumed when out of service, together with the losses in its generator transformers.

2.1.10 **Heat rate** The ratio of the energy content of the fuel used to the electrical energy produced over a given period; it can be referred to the electricity generated (gross) or the electricity supplied (net). The reciprocal of the heat rate expressed as a percentage is the *thermal efficiency* of the power station. In the case of the "heat rate" the units should be stated; in the case of the "thermal efficiency" the energy content of the fuel and the electrical energy produced must be expressed in the same unit.

2.2 Electricity Transmission and Distribution

2.2.1 **Electrical installation** Civil engineering works, buildings, machines, apparatus, lines and associated equipment together forming an integrated unit for the generation, conversion, transformation, transmission, distribution, storage or utilisation of electrical energy.

Industrie électrique

2.1 Production d'électricité

2.1.1 **Centrale thermique classique** Installation dans laquelle l'énergie chimique contenue dans des combustibles fossiles solides, liquides ou gazeux est transformée en énergie électrique.

2.1.2 **Centrale nucléaire** Installation dans laquelle l'énergie libérée à partir de combustibles nucléaires est transformée en énergie électrique (voir 7.1.1)

2.1.3 **Centrale hydraulique ou hydro-électrique** Installation dans laquelle l'énergie potentielle de gravité de l'eau est transformée en énergie électrique.

2.1.4 **Centrale de base** Centrale qui est utilisée essentiellement pour couvrir la charge de base.

2.1.5 **Centrale de pointe** Centrale qui est utilisée essentiellement pour participer à la couverture de la charge de pointe.

2.1.6 **Refroidissement à circuit ouvert** Procédé qui consiste à prélever de l'eau d'un plan d'eau ou d'un cours d'eau et à la rejeter, après passage et échauffement à travers les condenseurs de la centrale dans le plan d'eau ou le cours d'eau.

2.1.7 **Refroidissement par réfrigérant atmosphérique humide** Procédé qui consiste à faire passer l'eau de refroidissement, qui s'est échauffée à travers les condenseurs, dans les tours de réfrigération placées en aval où l'eau cède sa chaleur dans l'atmosphère, principalement par évaporation, puis est recyclée vers les condenseurs.

2.1.8 **Refroidissement par réfrigérant atmosphérique sec** Procédé qui consiste à dissiper dans l'atmosphère la chaleur provenant des condenseurs, uniquement par convection dans des tours de réfrigération.

2.1.9 **Consommation des auxiliaires (de la centrale)** Énergie électrique consommée par les installations auxiliaires et annexes, y compris pendant l'arrêt des installations principales, et en y ajoutant les pertes des transformateurs principaux.

2.1.10 **Consommation spécifique de chaleur** La consommation spécifique moyenne de chaleur pendant un intervalle de temps déterminé est le quotient de l'équivalent calorifique de combustible consommé par la quantité d'énergie électrique produite pendant l'intervalle de temps considéré. Cette consommation peut être brute ou nette comme l'énergie produite.

2.2 Transport et Distribution d'électricité

2.2.1 **Installation électrique** Ensemble des ouvrages de génie civil, de bâtiments de machines, d'appareils, de lignes et d'accessoires servant à la production, la conversion, la transformation, le transport, la distribution, l'utilisation d'énergie électrique. Se dit aussi du seul ensemble des machines électriques, du matériel électrique et des circuits électriques.

Elektrizitätswirtschaft

2.1 Elektroenergieerzeugung

2.1.1 Verbrennungskraftwerk Ein Verbrennungskraftwerk ist ein Kraftwerk, in dem die in festen, flüssigen und gasförmigen Brennstoffen enthaltene chemische Energie in elektrische Energie umgewandelt wird.

2.1.2 Kernkraftwerk Ein Kernkraftwerk ist ein Kraftwerk, in dem die aus nuklearen Brennstoffen freigesetzte Energie in elektrische Energie umgewandelt wird. (Siehe 7.1.1)

2.1.3 Wasserkraftwerk Ein Wasserkraftwerk ist eine Anlage zur Umwandlung der potentiellen Energie des Wassers in elektrische Energie.

2.1.4 Grundlastkraftwerk Ein Grundlastkraftwerk ist ein Kraftwerk, das vorwiegend zur Deckung der Grundlast eingesetzt wird.

2.1.5 Spitzenlastkraftwerk Ein Spitzenlastkraftwerk ist ein Kraftwerk, das vorwiegend zur Deckung der Spitzenlast dient.

2.1.6 Frischwasserkühlung (Durchlaufkühlung) Die Frischwasserkühlung ist ein Kühlverfahren, bei dem Wasser aus einem Gewässer entnommen und nach Durchströmen der Kondensatoren erwärmt in das Gewässer zurückgeleitet wird.

2.1.7 Kühlung mit Nasskühlturm Die Kühlung mit Nasskühlturm ist ein Kühlverfahren, bei dem das Kühlwasser im Kreislauf über den Kondensator und die nachgeschalteten Kühltürme geleitet wird, wo es die aufgenommene Kondensatorwärme vor allem durch Verdunstung an die Luft abgibt, bevor es zum Turbinenkondensator zurückläuft.

2.1.8 Kühlung mit Trockenkühlturm Die Kühlung mit Trockenkühlturm ist ein Kühlverfahren, bei dem die Abführung der Kondensationswärme des Turbinenabdampfes an die Umgebung über Kühltürme allein durch konvektiven Wärmeübergang erfolgt.

2.1.9 Kraftwerkseigenverbrauch Der Kraftwerkseigenverbrauch eines Kraftwerkblockes oder Kraftwerkes ist die elektrische Energie, die in den Neben- und Hilfsanlagen - auch bei Stillstand der Anlagen - verbraucht wird, zuzüglich der Verluste der Maschinentransformatoren.

2.1.10 Spezifischer Brennstoffwärmeverbrauch Der spez. Brennstoffwärmeverbrauch ist der Quotient aus dem Energieinhalt des verbrauchten Brennstoffs und der erzeugten elektrischen Energie. Er kann auf die Brutto- oder Nettoarbeit bezogen werden.

2.2 Elektroenergieübertragung und -verteilung

2.2.1 Elektrische Anlage Eine elektrische Anlage ist die Gesamtheit der zu einer Einheit zusammengeschlossenen Einrichtungen, die zur Erfüllung einer bestimmten Funktion, wie Erzeugung, Uebertragung, Verteilung, Speicherung oder Umwandlung von elektrischer Energie dient.

Industria Electrica

2.1 Producción de Electricidad

2.1.1 Central térmica convencional Instalación en la que la energía química, contenida en combustibles fósiles, sólidos, líquidos o gaseosos, es transformada en energía eléctrica.

2.1.2 Central nuclear Instalación en la que la energía liberada, a partir de combustibles nucleares, es transformada en energía eléctrica.(véase asimismo 7.1.1.)

2.1.3 Central hidráulica o hidroeléctrica Instalación en la que la energía potencial de gravedad del agua es transformada en energía eléctrica.

2.1.4 Central de base Central que, principalmente, se utiliza para cubrir la carga de base.

2.1.5 Central de puntas Central que se utiliza principalmente para cubrir las cargas de puntas.

2.1.6 Refrigeración en circuito abierto Procedimiento que consiste en tomar el agua de un embalse o una corriente de agua y volverla a arrojar caliente al mismo embalse o corriente después de pasar por los condensadores de la central.

2.1.7 Refrigeración con torre de refrigeración húmeda Procedimiento por el que el agua de refrigeración que se ha calentado a través de los condensadores, se hace pasar por torres de refrigeración situadas aguas abajo para ceder su calor a la atmósfera principalmente por evaporción, con reciclado posterior.

2.1.8 Refrigeración con torre de refrigeración seca Procedimiento por el que el calor procedente de los condensadores de una central se disipa a la atmósfera exclusivamente por convección en torres de refrigeración.

2.1.9 Consumo propio de la central Energía eléctrica consumida por una central en sus servicios auxiliares, incluyendo el consumo cuando está fuera de servicio, así como las pérdidas de los transformadores principales.

2.1.10 Consumo específico de calor El consumo específico medio de calor en un intervalo de tiempo determinado es el cociente entre el equivalente calorífico del combustible consumido y la cantidad de energía eléctrica producida en el intervalo de tiempo considerado. Lo mismo que la energía producida este consumo puede ser bruto o neto.

2.2 Transporte y Distribución de Electricidad

2.2.1 Instalación eléctrica Conjunto de obras de ingeniería, edificios, máquinas, aparatos, líneas y accesorios, que sirven para la producción, conversión, transformación, transporte, distribución, utilización de energía eléctrica. Se aplica también esta denominación a un solo conjunto de máquinas eléctricas, de material eléctrico o de circuitos eléctricos.

2

2.2.2 **Electric line** A generic term for a set of conductors, with insulation and accessories, used for the transmission or distribution of electrical energy.

2.2.3 **Overhead line/cable** An electric line situated above ground usually with the conductors supported on insulators and appropriate supports. The term would include works and fittings associated with the line.

2.2.4. **Underground/submarine line/cable** An electric line situated in the ground/under water. The term would include works and fittings associated with the line.

2.2.5 **Single circuit line** A line having only one circuit.

2.2.6 **Multiple circuit line** A line comprising several circuits.

2.2.7 **Electric circuit** An arrangement of bodies or media through which a current can flow.

2.2.8 **Circuit length** The average of the actual lengths of the lines of a circuit (taking account of variations in elevation and catenary dip).

2.2.9 **(Transmission or distribution) route/right of way** The terrain required for running an overhead line or an underground line.

2.2.10 **(Transmission or distribution) route length** The distance between the end points of an overhead line or underground line, horizontally projected and measured along the route axis.

2.2.11 **Switching station** An electrical installation for the selective connection and disconnection of the lines of a system/network and of consumer installations by means of switchgear.

2.2.12 **Transforming station** A substation which includes transformers for transferring electricity between systems operating at different voltage levels.

2.2.13 **Distribution substation; HV/LV transforming station** A transforming station between high and low voltage systems/networks.

2.2.14 **Converter station** An installation for converting current of one form into another or for converting one frequency into another.

2.2.15 **Rectifier station** An installation for converting single or multi-phase alternating current into direct current.

2.2.16 **Inverter station** An installation for converting direct current into single or multi-phase alternating current.

2.2.17 **Network; system** A grouping of lines and of other electrical equipment connected for the purpose of conveying electricity from generating stations to the ultimate consumer.

2.2.18 **Interconnected or interconnecting network/system** A network that can be so regulated in its overall performance, both nationally and/or internationally that it enables electricity demand to be met with electricity generation optimally, both as regards economics and reliability.

2.2.19 **Transmission network/system** A system of transmission lines serving for the super-regional transport of electricity and feeding to subsidiary systems.

2.2.2 **Ligne** Ensemble des conducteurs, d'isolants et d'accessoires destinés au transport ou à la distribution de l'énergie électrique.

2.2.3 **Ligne aérienne** Ligne dont les conducteurs sont maintenus au-dessus du sol, généralement au moyen d'isolateurs et de supports appropriés (pylônes, massifs . .) y compris les dispositifs accessoires (prise à la terre . .).
N.B.: Une ligne aérienne peut également être sous forme de câble.

2.2.4 **Câble (ou ligne) souterrain** Ligne dont les conducteurs sont posés dans le sol (ou sous l'eau), y compris les dispositifs accessoires.

2.2.5 **Ligne simple** Ligne ne comportant qu'un seul circuit électrique.

2.2.6 **Ligne multiple** Ligne comportant plusieurs circuits électriques.

2.2.7 **Circuit de ligne électrique** Ensemble de conducteurs formant un système indissociable électriquement et transportant l'énergie électrique.

2.2.8 **Longueur d'un circuit de ligne électrique** La longueur d'un circuit de ligne électrique est la moyenne des longueurs réelles des conducteurs de ce circuit électrique (tient compte donc des différences de niveau et de flèches).

2.2.9 **Tracé** Terrain couvert ou suivi par le cheminement d'une ligne aérienne ou d'un câble.

2.2.10 **Longueur du tracé ou longueur géographique d'une ligne ou d'un circuit** Longueur de la projection horizontale sur le tracé de la distance entre les extrémités d'une ligne aérienne ou d'un

2.2.11 **Poste de sectionnement (installation à haute tension)** Installation électrique dont la fonction est de connecter ou de déconnecter des lignes d'un réseau ou des points de livraison en antenne.

2.2.12 **Poste de transformation** Installation électrique dont la fonction est de transférer de l'énergie électrique entre des réseaux de tensions différente.

2.2.13 **Poste de transformation HT/BT** Poste de transformation entre réseaux haute tension et basse tension.

2.2.14 **Convertisseur** Installation dont la fonction est de convertir un type de courant en un autre ou une fréquence en une autre.

2.2.15 **Redresseur** Installation dont la fonction est de convertir du courant alternatif (monophasé ou polyphasé) en courant continu.

2.2.16 **Onduleur** Installation dont la fonction est de convertir du courant continu en courant alternatif.

2.2.17 **Réseau** Ensemble des lignes et ouvrages électriques connectés entre eux et intervenant dans l'acheminement de l'énergie électrique.

2.2.18 **Réseau d'interconnexion** Réseau qui, sur le plan national ou international réalise la liaison assurant des mouvements d'énergie entre réseaux, entre centrales ou entre réseaux et centrales. Il sert à augmenter la rentabilité et la fiabilité de l'alimentation en énergie électrique.

2.2.19 **Réseau de transport** Réseau servant au transport interrégional de l'énergie électrique vers les réseaux aval.

2.2.2 Leitung Eine Leitung ist eine elektrotechnische Einrichtung zur Fortleitung von elektrischer Energie.

2.2.3 Freileitung Eine Freileitung ist eine Leitung, die aus der Gesamtheit von freigespannten Leiterseilen, Tragwerken samt Fundamenten, Erdungen, Isolatoren, Zubehör und Armaturen besteht.
Eine Freileitung kann auch aus Kabelleitern bestehen.

2.2.4 Kabelleitung Eine Kabelleitung ist eine Leitung, die im Erdreich oder Wasser verlegt ist und aus der Gesamtheit von isolierten Leitern, Muffen, Endverschlüssen und Zubehör besteht.

2.2.5 Einfachleitung Eine Einfachleitung ist eine Leitung, die aus einem Stromkreis besteht.

2.2.6 Mehrfachleitung Eine Mehrfachleitung ist eine Leitung, die aus mehreren Stromkreisen besteht.

2.2.7 Stromkreis Ein Stromkreis ist die Gesamtheit von Leitern, die eine elektrisch-physikalische nicht trennbare Einheit bilden.

2.2.8 Stromkreislänge Die Stromkreislänge ist das Mittel aus den tatsächlichen Längen der Leiter eines Stromkreises (unter Berücksichtigung von Höhenunterschied und Durchhang).

2.2.9 Trasse Die Trasse ist der für die Führung einer Freileitung oder Kabelleitung erforderliche Geländebereich.

2.2.10 Trassenlängen Die Trassenlänge ist die auf die Horizontale projizierte, in Trassenachse gemessene Entfernung zwischen den Endpunkten einer Freileitung oder Kabelleitung.

2.2.11 Schaltanlage Eine Schaltanlage ist eine elektrische Anlage zur wahlweisen Verbindung und Auftrennung von Leitungen eines Netzes und von Abnehmereinrichtungen mittels Schalter.

2.2.12 Umspannanlage Eine Umspannanlage sit eine elektrische Anlage zur Uebertragung von elektrischer Energie zwischen Netzen mit unterschiedlichen Spannungsebenen.

2.2.13 Ortsnetzstation (Transformatorenstation) Eine Ortsnetzstation (Transformatorenstation) ist eine Umspannanlage zwischen Hoch- und Niederspannungsnetzen.

2.2.14 Umrichteranlage Eine Umrichteranlage ist eine Anlage zur Umformung einer Stromart oder Frequenz in eine andere.

2.2.15 Gleichrichteranlage Eine Gleichrichteranlage (Gleichrichterstation) ist eine Umrichteranlage zur Umformung von ein- oder mehrphasigem Wechselstrom in Gleichstrom.

2.2.16 Wechselrichteranlage Eine Wechselrichteranlage (Wechselrichterstation) ist eine Umrichteranlage zur Umformung von Gleichstrom in ein- oder mehrphasigen Wechselstrom.

2.2.17 Netz Ein Netz ist die Gesamtheit aller miteinander verbundenen Leitungen, Schalt- und Umspannanlagen.

2.2.18 Verbundnetz Ein Verbundnetz ist ein übergeordnetes Netz, das national und/oder international die Schwerpunkte der Erzeugung und des Verbrauchs elektrischer Energie miteinander verbindet und der Verbesserung der Writschaftlichkeit und der Zuverlässigkeit der Versorgung dient.

2.2.19 Uebertragungsnetz Ein Uebertragungsnetz dient dem überregionalen Transport elektrischer Energie zu nachgeordneten Netzen.

2.2.2 Línea Conjunto de conductores, aislantes y accesorios destinados al transporte o a la distribución de la energía eléctrica.

2.2.3 Línea aérea Línea cuyos conductores se mantienen por encima del suelo por medio, generalmente, de aisladores y de soportes apropiados (torres, macizos. . . .) incluídos los dispositivos accesorios (puesta a tierra. . . .).
Nota: Una línea aérea puede estar formada también por cables.

2.2.4 Cable subterráneo Línea cuyos conductores, incluídos los dispositivos accesorios, se colocan bajo tierra (o bajo el agua).

2.2.5 Línea de circuito simple Línea provista de un solo circuito eléctrico.

2.2.6 Línea de circuito múltiple Línea provista de varios circuitos eléctricos.

2.2.7 Circuito eléctrico Conjunto de conductores que forman un sistema eléctricamente indisociable y que transportan la energía elétrica.

2.2.8 Longitud de un circuito eléctrico Es la media de las longitudes reales de los conductores de dicho circuito elétrico (habida cuenta de las diferencias de nivel y de las flechas).

2.2.9 Trazado Es la faja de terreno requerido para tender una línea aérea o subterránea.

2.2.10 Longitud del trazado (en transporte o distribución) Es la distancia entre extremos de una línea aérea o subterránea proyectada horizontalmente y medida por el eje de la traza.

2.2.11 Subestación de seccionamiento (instalación de alta tensión) Es una instalación eléctrica para la conexión y desconexión selectiva de las lineas de una red o de centros de suministro, por medio de interruptores.

2.2.12 Subestación de transformación Instalación eléctrica cuya finalidad es la transferencia de energía eléctrica entre redes a tensiones diferentes.

2.2.13 Subestación de transformacion de AT/BT Subestación de transformación entre redes de alta tensión y baja tensión.

2.2.14 Convertidor Es una instalación para convertir la corriente de una forma a otra o para convertir una frecuencia a otra.

2.2.15 Rectificador Instalción cuya finalidad es convertir una corriente alterna (monofásica o polifásica) en corriente contínua.

2.2.16 Ondulador Instalación cuya finalidad es convertir la corriente contínua en corriente alterna.

2.2.17 Red Conjunto de líneas y obras eléctricas conectadas entre sí para la conducción de la energía eléctrica.

2.2.18 Red de interconexión Red que, a nivel nacional o internacional, realiza la conexión asegurando las transferencias de energía entre redes, entre centrales, o entre redes y centrales. Sirve para aumentar la rentabilidad y la fiabilidad del suministro de energía eléctrica.

2.2.19 Red de transporte Red que sirve para el transporte internacional de la energía eléctrica, bacia las redes situadas aguas abajo.

2

2.2.20 Distribution network/system A system of distribution lines serving for the regional and local distribution of electrical energy.

2.2.21 Radial system A system or part of a system which is wholly or mainly composed of radial circuits and hence points to be supplied do not have a supply available to them in more than one direction.

2.2.22 Ringed network/system A network or part of a network which is wholly or mainly composed of ring circuits, all or most of which individually emanate from and terminate at the same source of supply.

2.2.23 Meshed network/system A network or part of a network which is wholly or mainly composed of ring circuits all or most of which emanate from and terminate at different sources of supply, or any more complex system of multiple ring circuits with multiple supply sources.

2.2.24 High voltage A voltage equal to or higher than a specified voltage that may vary legally from one country to another (e.g. in most European countries it now applies to voltages above 1000 V between conductors).

2.2.25 Low voltage A voltage equal to or lower than a specified voltage that may vary legally from one country to another (e.g. in most countries it applied to voltages of 1000 V a.c. or below between conductors).

2.2.26 Rated voltage The voltage used in the specification of a machine, plant, network or apparatus and from which the test conditions and the voltage limits for the use of the machine etc. are calculated.

2.2.27 Operating voltage The voltage at any moment across two line wires of machines or apparatus in operation.

2.2.28 Transmission capacity/capability The highest permissible continuous loading of the transmission equipment with respect to heating, stability and voltage drop.

2.2.29 Interconnection The connection, by one or more lines, between two or more systems or parts of systems, and the equipment for such connection.

2.2.30 Supply terminals; delivery/terminal point The point in a system/network at which a purchaser/consumer contractually receives electrical energy.

2.2.31 Network/system internal consumption Such consumption of electrical energy by ancillary equipment as is required for the operation of the network/system.

2.2.32 Network/system losses; transmission and distribution losses The energy losses occurring in transmission and distribution in a specific network/system.

2.3 Capacity and Generation

2.3.1 Installed capacity; gross installed capacity The capacity measured at the output terminals of all sets in the station; it includes power taken for the station's internal load.

2.3.2 Maximum output capacity; net output capacity; output capacity The capacity measured at the point of outlet to the network.

2.3.3 Power station internal load; station service load; auxiliaries load The electrical capacity of a power station or power station set, that is required for its auxiliary plant, together with the capacity represented by the losses in its generator transformers.

2.2.20 Réseau de distribution Réseau servant à la distribution de l'énergie électrique à l'intérieur d'une région délimitée.

2.2.21 Réseau radial Réseau ou partie d'un réseau, entièrement ou principalement constitué de lignes en antenne.

2.2.22 Réseau bouclé Réseau ou partie de réseau, constitué entièrement ou principalement par des boucles, dont la totalité ou la plupart sont individuellement raccordées, à leurs extrêmités, à la même source d'alimentation.

2.2.23 Réseau maillé Réseau, ou partie d'un réseau, entièrement ou principalement et constitué de boucles raccordées, à leurs extrêmités, à des sources d'alimentation différentes ou tout ensemble plus complexe constitué de boucles multiples ou plusieurs sources d'alimentation.

2.2.24 Haute tension La haute tension est une tension dont la valeur entre phases est égale ou supérieure à une tension spécifique qui varie d'un pays à l'autre et en général supérieure à 1 000 Volts.

2.2.25 Basse tension La basse tension est une tension dont la valeur entre phases est égale ou inférieure à une tension spécifique qui varie d'un pays à l'autre et en général inférieur à 1 000 Volts.

2.2.26 Tension nominale Tension qui figure dans la désignation d'une machine ou d'un appareil et d'après laquelle sont déterminées les conditions d'essai et les tensions limites d'utilisation.

2.2.27 Tension d'exploitation Tension sous laquelle sont en service des installations électriques (production, transport...).

2.2.28 Capacité de transport La capacité de transport d'un moyen d'exploitation est la charge maximale admissible en permanence en tenant compte de l'échauffement, de la stabilité et de la chute de tension.

2.2.29 Interconnexion Connexion, par une ou plusieurs lignes, entre deux ou plusieurs réseaux.

2.2.30 Point d'échange Point du réseau ou l'énergie électrique est transmise à un autre réseau.

2.2.31 Consommation propre du réseau Consommation d'énergie électrique des installations auxiliers et annexes nécessaires au fonctionnement du réseau.

2.2.32 Perte de réseau Pertes d'énergie électrique qui se produisent dans le réseau considéré, du fait du transport et de la distribution d'énergie électrique:

2.3 Puissance et Energie

2.3.1 Puissance brute Puissance électrique mesurée aux bornes du générateur.

2.3.2 Puissance nette Puissance électrique mesurée à la sortie de la centrale.

2.3.3 Puissance absorbée par les auxiliaires Puissance électrique absorbée par les installations ou services auxiliaires d'une tranche ou d'une centrale, en y ajoutant la puissance dissipée dans les transformateurs de centrale (perte des transformateurs principaux).

2.2.20 Verteilungsnetz Ein Verteilungsnetz dient innerhalb einer begrenzten Region der Verteilung elektrischer Energie.

2.2.21 Strahlennetz Ein Strahlennetz ist ein Netz, in dessen Leitungen nur von einem Ende aus eingespeist werden kann.

2.2.22 Ringnetz Ein Ringnetz ist ein Netz, dessen Leitungen von einem Speisepunkt ausgehend wieder zu ihm zurückführen.

2.2.23 Maschennetz Ein Maschennetz ist ein Netz, dessen Leitungen an mehreren Knotenpunkten zusammengeführt sind. Es ist im allgemeinen durch vielfache Leitungsverbindungen und mehrere Speisepunkte gekennzeichnet.

2.2.24 Hochspannung Hochspannung ist eine Spannung mit einem Wert der Leix-Leiter-Spannung, der von Land zu Land verschieden, meistens ≥ 1000 V ist.

2.2.25 Niederspannung (Wechselspannung) Niederspannung ist eine Spannung mit einem Wert der Leiter-Leiter-Spannung, der von Land zu Land verschieden, meistens ≤ 1000 V ist.

2.2.26 Nennspannung Nennspannung ist diejenige Spannung, nach der ein Betriebsmittel, eine Anlage oder ein Netz benannt wird und auf die bestimmte Betriebseigenschaften bezogen werden.

2.2.27 Betriebsspannung Die Betriebsspannung ist die jeweils örtlich zwischen zwei Leitern herrschende Spannung von Betriebsmitteln oder Anlageteilen.

2.2.28 Uebertragungsfähigkeit Die Uebertragungsfähigkeit eines Betriebsmittels ist die höchste zuässige Dauerbelastung im Hinblick auf Erwärmung, Stabilität und Spannungsabfall.

2.2.29 Netzkupplung Eine Netzkupplung ist eine elektrische Anlage, die zur Kupplung von Netzen oder Netzteilen dient.

2.2.30 Uebergabestelle Eine Uebergabestelle ist eine Stelle im Netz, an der elektrische Energie vereinbarungsgemäss einem anderen übergeben wird.

2.2.31 Eigenverbrauch des Netzes Der Eigenverbrauch eines Netzes ist der Verbrauch an elektrischer Energie von Hilfs- und Nebenanlagen, die für den Betrieb des Netzes notwendig sind.

2.2.32 Netzverluste (Uebertragungsverluste) Netzverluste (Uebertragungsverluste) sind die Energieverluste, die im betrachteten Netz für Uebertragung und Verteilung auftreten.

2.3 Leistung und Erzeugung

2.3.1 Bruttoleistung Die Bruttoleistung ist die an den Generatorklemmen gemessene Leistung.

2.3.2 Nettoleistung Die Nettoleistung ist die für das Netz nutzbare Leistung eines Kraftwerklockes oder Kraftwerkes.

2.3.3 Eigenbedarfsleistung Die Eigenbedarfsleistung eines Kraftwerkblockes oder eines Kraftwerkes ist die elektrische Leistung, die in seinen Neben- und Hilfsanlagen benötigt wird, zuzüglich der Verlustleistung der Maschinentransformatoren.

2.2.20 Red de distribución Red que sirve para la distribución de energía en el interior de una región o zona determinada.

2.2.21 Red radial Red o parte de una red, total o parcialmente constituída por líneas que parten de un centro.

2.2.22 Red de anillos Red, o parte de una red, constituída total o principalmente por anillos cuya totalidad o mayoría están conectados individualmente, por sus extremos, a la misma fuente de alimentación.

2.2.23 Red mallada Red o parte de una red, total o principalmente compuesta por anillos cuya totalidad o mayoría están conectados por sus extremos a fuentes de alimentación diferentes, o cualquier conjunto más complejo constituído por anillos múltiples y varias fuentes de alimentación.

2.2.24 Alta tensión Es una tensión cuyo valor entre fases es igual o superior a una tensión especificada que varía de un país a otro. En numerosos países europeos la alta tensión se define como una tensión estrictamente superior a 1 kV.

2.2.25 Baja tensión Es una tensión cuyo valor entre fases es igual o inferior a una tensión especificada que varía de un país a otro. En la mayoría de los países se aplica a tensiones de 1000 V o menos en corriente alterna entre fases.

2.2.26 Tensión nominal Tensión que figura en las especificaciones de una máquina o de un aparato, a partir de la cual se determinan las condiciones de prueba y las tensiones límites de utilización de esta máquina o de este aparato.

2.2.27 Tensión de explotación Tensión bajo la que se encuentran en servicio las instalaciones eléctricas (producción, transporte, etc. . . . .).

2.2.28 Capacidad de transporte De un medio de explotación, es la carga máxima admisible con carácter contínuo habida cuenta del calentamiento, la estabilidad y la caída de tensión.

2.2.29 Interconexión Conexión entre dos o más redes mediante una o varias líneas.

2.2.30 Punto de intercambio Punto de una red por el que se transmite energía eléctrica a otra red.

2.2.31 Consumo interno de una red Consumo de energía eléctrica de las instalaciones auxiliares anexas necesarias para el funcionamiento de la red.

2.2.32 Pérdidas de una red Son las pérdidas de energía eléctrica que se producen por su transporte y distribución en la red considerada.

2.3 Potencia y Producción

2.3.1 Potencia bruta Es la potencia eléctrica medida en los terminales del genrador.

2.3.2 Potencia neta Es la potencia eléctrica medida a la salida de la central.

2.3.3 Potencia para servicios auxiliares Potencia eléctrica absorbida por las instalaciones o servicios auxiliares de una central o conjunto de centrales junto con la potencia perdida en los transformadores de la central (pérdidas de los transformadores principales).

2.3.4 Maximum capacity; maximum electric capacity In the case of a thermal unit or station, the maximum power that could be produced under continuous operation with all plant running and with adequate fuel stocks of normal quality. In the case of a hydro-electric installation, the maximum power that could be produced throughout a given period of operation with all plant running and with flow and head at their optimum values.

2.3.5 Available capacity; available power At any given moment, the maximum power at which the station or unit can be operated for a given period under the prevailing conditions assuming unlimited transmission facilities.

2.3.6 Power produced, utilised capacity; operating capacity The actual capacity operated. In principle it is measured as an instantaneous value and must be referred to a time; however, by convention it may be derived from the energy produced during a certain period which for statistics it is necessary to define (the ratio of the electricity produced to the operating period). The power produced may be gross or net.

2.3.7 Reserve capacity Cold stand-by (in exceptional cases), hot stand-by and spinning reserve capacities that serve to meet any difference between the anticipated capacity demand and the capacity demand actually occurring.

2.3.8 Minimum stable generation/capacity The lowest capacity at which a station can be operated without technical difficulty.

2.3.9 Optimum capacity The capacity at which a system or a station has its highest efficiency.

2.3.10 Maximum power produced The maximum value of output or load which can be maintained for a specified period.

2.3.11 Minimum capacity The lowest capacity in a given period.

2.3.12 Firm capacity The capacity which can be made available, whose reliability for the supply system is specified and determined in advance.

2.3.13 Electricity generated The electricity produced at the generator terminals.

2.3.14 Electricity supplied The useful electricity supplied to the network

2.3.15 Input to network The sum of the electricity supplied by the electricity generators of the network and supplies from other sources.

2.4 Operation of the Electricity System

2.4.1 Control room A room in which control boards are installed.

2.4.2 System control centre (center) The appropriate centre (center) for switching or directing the switching of the lines of a network/system.

2.4.3 Load dispatching centre (center) The appropriate centre (center) for switching or directing the switching of power stations on line and for load changing. In general the load dispatching centre (center) and the system control centre (center) are one and the same in the case of centrally controlled systems/networks.

2.3.4 Puissance électrique maximale possible Puissance électrique maximale réalisable par une tranche ou une centrale pendant un temps de fonctionnement déterminé, la totalité de ses installations étant supposée entièrement en état de marche, les conditions d'alimentation (combustible ou eau) étant optimales.

2.3.5 Puissance électrique disponible Puissance électrique maximale réalisable par une tranche ou une centrale pendant un temps de fonctionnement déterminé et dans les conditions réelles où elle se trouve à cet instant, à l'exclusion toutefois des possibilités d'évacuation de l'énergie électrique produite, qui sont supposées illimitées.

2.3.6 Puissance électrique produite Puissance effectivement réalisée. Elle est mesurée, en principe, d'une manière instantanée en étant complétée par l'indication du moment. A défaut, la puissance produite peut être conventionellement déterminée en partant de l'énergie électrique produite pendant un certain intervalle de temps (quotient production par durée).

2.3.7 Puissance de réserve Puissance servant à compenser les écarts entre les charges prévues et les charges réalisées.

2.3.8 Puissance minimale technique (minimum technique) Plus faible puissance à laquelle peut fonctionner l'installation dans des conditions techniques correctes.

2.3.9 Puissance optimale Puissance avec laquelle une installation a son meilleur rendement.

2.3.10 Puissance maximale produite Maximum constaté, pendant un intervalle de temps déterminé, de la puissance électrique produite par une installation.

2.3.11 Puissance minimale C'est la puissance minimale pendant une période de temps donnée.

2.3.12 Puissance garantie Puissance qui peut être mise à disposition avec une fiabilité prédéterminée.

2.3.13 Energie produite brute Énergie électrique mesurée aux bornes du générateur.

2.3.14 Energie produite nette Énergie électrique mesurée à la sortie de la centrale.

2.3.15 Energie fournie au réseau Ensemble de l'énergie électrique fournie au réseau, c'est-à-dire somme de l'énergie produite nette par ses propres centrales et de l'énergie d'autre provenance.

2.4 Exploitation

2.4.1 Salle de commande Endroit où sont groupés les tableaux de commande des installations.

2.4.2 Poste de commande Organe dont la fonction est de diriger l'exploitation des lignes d'un réseau.

2.4.3 Centre de conduite (Dispatching) Installation dont la fonction est de mettre en oeuvre les centrales et de répartir les charges. En général, il commande également la commutation des réseaux directement concernés.

2.3.4 Engpassleistung Die Engpassleistung eines Kraftwerblockes oder eines Kraftwerkes ist die höchste Dauerleistung, die unter optimalen Bedingungen (Brennstoff oder Wasser) ausfahrbar ist.

2.3.5 Verfügbare Leistung Die verfügbare Leistung ist die höchste Leistung, die zur jeweiligen Zeit unter Berücksichtigung aller technischen und betrieblichen Verhältnisse (ohne Netzeinflüsse) erreicht werden kann.

2.3.6 Verfügte Leistung (Betriebsleistung) Die verfügte Leistung (Betriebsleistung) ist die tatsächlich gefahrene Leistung. Sie fällt normalerweise als Momentanwert an und ist mit Zeitangabe zu versehen. Anderenfalls ist die verfügte Leistung (Betriebsleistung) als Mittelwert über die Betriebszeit aufzufassen, der sich als Quotient aus Erzeugung und Betriebszeit ergibt.

2.3.7 Reserveleistung Die Reserveleistung dient zum Ausgleich der Abweichungen in der Leistungsbilanz zwischen den erwarteten und den eintretenden Verhältnissen.

2.3.8 Technische Mindestleistung Die technische Mindestleistung ist die niedrigste Leistung, mit der eine Anlage technisch einwandfrei betrieben werden kann.

2.3.9 Bestleistung Die Bestleistung ist die Leistung, bei der ein Aggregat oder eine Anlage den höchsten Wirkungsgrad hat.

2.3.10 Höchstleistung Die Höchstleistung ist die höchste erreichte Leistung in einer bestimmten Zeitspanne.

2.3.11 Niedrigstleistung Die Niedrigstleistung ist die niedrigste erreichte Leistung in einer bestimmten Zeitspanne.

2.3.12 Gesicherte Leistung Die gesicherte Leistung ist die Leistung, die mit einer bestimmten vorzugebenden Versorgungszuverlässigkeit bereitgestellt werden kann.

2.3.13 Bruttoerzeugung Die Bruttoerzeugung ist die an den Generatorklemmen gemessene Energie.

2.3.14 Nettoerzeugung Die Nettoerzeugung ist die für das Netz nutzbare Energie.

2.3.15 Netzeinspeisung Die Netzeinspeisung ist die elektrische Energie, die als Summe von Nettoerzeugung und Bezug in ein Netz eingespeist wird.

2.4 Betrieb des Elektroenergiesystems

2.4.1 Schaltwarte Die Schaltwarte ist die Stelle, an der die Ueberwachungs- und Steuereinrichtungen von Schaltanlagen oder Netzen zusammengefasst sind.

2.4.2 Schaltleitung Die Schaltleitung ist die für die Schaltung von Leitungen eines Netzes zuständige Stelle.

2.4.3 Lastverteilung Die Lastverteilung ist die für den Einsatz der Kraftwerke zuständige Stelle. Im allgemeinen stellt die Lastverteilung auch die Schaltleitung für übergeordnete Netze dar.

2.3.4 Máxima potencia eléctrica posible Potencia máxima que puede ser realizada por un sistema o una central durante un tiempo determinado de funcionamiento, suponiendo en servicio la totalidad de sus instalaciones y en condiciones óptimas de alimentación (combustible o agua).

2.3.5 Potencia eléctrica disponible Potencia eléctrica máxima que puede ser realizada por un sistema o una central durante un tiempo determinado de funcionamiento y en las condiciones reales que se encuentra en ese - instante excluyendo, sin embargo, cualquier limitación en la salida de la energía eléctrica producida, que se supone ilimitada.

2.3.6 Potencia eléctrica utilizada Potencia realmente empleada. En principio se mide como un valor instantáneo y debe indicarse el momento a que se refiere. En su defecto puede expresarse convencionalmente la potencia utilizada, a partir de la energía eléctrica producida durante un intervalo de tiempo determinado (cociente entre energía producida y tiempo).

2.3.7 Potencia de reserva Potencia que sirve para compensar las desviaciones entre las cargas previstas y las reales.

2.3.8 Potencia técnica mínima (mínimo técnico) La potencia más débil a la que puede funcionar una central en condiciones tcnicas correctas.

2.3.9 Potencia óptima Potencia a la que un sistema o una central tiene su mejor rendimiento.

2.3.10 Potencia máxima producida Máximo comprobado, durante un intervalo de tiempo determinado, de la potencia eléctrica alcanzada por una instalación.

2.3.11 Potencia mínima Es la potencia más baja en un período dado de tiempo.

2.3.12 Potencia garantizada Potencia que puede ser puesta a disposición con una fiabilidad determinada de antemano.

2.3.13 Energía bruta producida Energía eléctrica medida en los terminales del generador.

2.3.14 Energía neta producida Energía eléctrica medida a la salida de la central.

2.3.15 Energía entregada a la red Conjunto de la energía suministrada a la red, es decir, suma de la energía neta producida más la recibida de otras fuentes.

2.4 Explotación

2.4.1 Sala de mando Lugar donde se agrupan los cuadros de mando de las instalaciones.

2.4.2 Centro de maniobra Es el centro apropiado para dirigir la explotación de las líneas de una red.

2.4.3 Repartidor de cargas Instalación cuya función es la puesta en servicio de las centrales y distribuir las cargas. En general, dirige igualmente la conmutación de las redes afectadas directamente.

2.4.4 Ripple control A method of load management control which involves connecting and disconnecting consumer groups, the necessary remote control being effected via the distribution network/system.

2.4.5 Demand curve; load curve A curve representing the changing values of output or load as a function of time.

2.4.6 Load-controlled consumer A consumer of electricity whose demand may be regulated in such a way that it contributes to flattening the load curve of the electricity supply network/system; an *interruptible consumer* is a particular case of a load-controlled consumer.

2.4.7 Chargeable demand The demand taken into account for calculating the charges to be billed.

2.4.4 Télécommande centralisée Méthode pour connecter ou déconnecter des groupes de consommateurs généralement par télécommande sur le réseau de distribution.

2.4.5 Diagramme de charge, Courbe de charge Représentation graphique de l'évolution de la charge en fonction du temps.

2.4.6 Consommateur modulable Consommateur d'énergie électrique dont on peut moduler la demande de puissance afin de le faire contribuer à la régularisation de la courbe de charge. Cas particulier : client interruptible.

2.4.7 Puissance de facturation Puissance prise en considération pour le calcul du prix facturé.

2.4.4 Rundsteuerung Die Rundsteuerung ist eine Methode, um Verbrauchergruppen zu- und abzuschalten, wobei die Fernsteuerbefehle in der Regel über das Verteilungsnetz übertragen werden.

2.4.5 Leistungsganglinie Die Leistungsganglinie ist eine graphische Darstellung des Leistungs- oder Lastverlaufs in Funktion der Zeit.

2.4.6 Regelbare Verbraucher Regelbare Verbraucher sind Verbraucher von elektrischer Energie, bei denen die Möglichkeit besteht, sie an der Vergleichmässigung der Belastungsganglinie des elektrischen Netzes zu beteiligen.

Spezialfall : vertraglich vereinbarter Lastabwurf.

2.4.7 Verrechnungsleistung Die Verrechnungsleistung ist die Leistung, die für die Berechnung dem Abnehmer gemäss Vertrag zur Bildung des Leistungspreises zugrunde gelegt wird.

2.4.4 Telemando centralizado Método de conectar y desconectar grupos consumidores, generalmente por telemando sobre la red de distribución.

2.4.5 Diagrama de carga, curva de carga Representación gráfica de la evolución de una carga en función del tiempo.

2.4.6 Abonado de consumo regulable Consumidor de energía electrica cuya demanda puede ser regulada de tal manera que contribuye a aplanar la curva de carga.

Caso particular: abonado interrumpible.

2.4.7 Potencia de facturación Potencia tomada en consideración para el cálculo del precio que se factura.

2

Section 3

Hydro-electricity - Water Power
Energie hydro-électrique (énergie hydraulique)
Wasserkraftwirtschaft (Hydroenergetik)
Energia Hidroelectrica (energía hidráulica)

3

Hydro-electricity - Water Power

3.1 General Terms

3.1.1 Water power; hydro-electric power; hydro-power The potential energy of waters

3.1.2 Hydro-electric power station A plant designed to convert the gravitational energy of waters into electrical energy.

3.1.3 Run-of-river power station A hydro-electric power station which has no significant regulating reservoir.

3.1.4 Power station with reservoir A hydro-electric power station associated with storage capacity to regulate the water supply to the turbines.

3.1.5 Pumped storage power station; pumped storage plant A power station with reservoir, the reservoir being filled exclusively or partially with the use of pumps.

3.1.6 Tidal power station A hydro-electric power station that exploits the head occurring between the level of the water in the sea, at low tide and that of the seawater impounded in a basin separated from the sea, at high tide.

3.2 Terms relating to Time

3.2.1 Water resources year; annual run-off A one-year period (i.e. a period of 12 consecutive calendar months) based on hydro-electric power considerations.

3.2.2 Mean year; normal year; average year A hypothetical year whose hydrological data are, for the purposes of water power, averages of a consecutive series of as many years as possible. The number of consecutive years upon which the mean year or normal year is based must be stated in each case.

3.2.3 Wet year; peak run-off year A (hypothetical) year in which the flow of water is greater than that of a hypothetical average year based on statistical criteria.

3.2.4 Dry year; minimum run-off year A (hypothetical) year in which the flow of water is less than that of a hypothetical average year based on statistical criteria.

3.2.5 Draw-off period of a reservoir; reservoir draw-down time The minimum period required to empty the reservoir from the highest to the lowest level normally allowable in use through the turbines of its own station, assuming the absence of all natural inflows.

3.2.6 Filling period of a reservoir The time required for filling the reservoir from the lowest to the highest level normally allowable in use, with constant supply flow equal to the characteristic mean corrected flow.

3.2.7 Filling period of a pumped storage reservoir; pumping time The time taken to fill the upper basin of a pumped storage reservoir from the lowest to the highest level normally allowable in use, assuming the absence of all natural inflows, when the pumps are working at full capacity.

Energie hydro-électrique (énergie hydraulique)

3.1 Termes généraux

3.1.1 Energie hydraulique Énergie potentielle des eaux.

3.1.2 Centrale hydraulique Installation dans laquelle l'énergie potentielle de gravité de l'eau est transformée en énergie électrique.

3.1.3 Centrale hydro-électrique au fil de l'eau Centrale hydraulique qui n'a pas de régulation sensible par réservoir.

3.1.4 Centrale hydro-électrique à réservoir Centrale hydro-électrique dont l'alimentation peut être régularisée grâce à un réservoir.

3.1.5 Aménagement hydraulique à accumulation par pompage Centrale hydro-électrique à réservoir, ce dernier étant rempli totalement ou partiellement au moyen de pompes.

3.1.6 Centrale marémotrice Centrale hydraulique utilisant la hauteur de chute entre la mer et un bassin qui en est séparé et utilisant les variations du niveau de la mer provoquées par le flux et le reflux.

3.2 Termes Relatifs au Temps

3.2.1 Année hydrologique Période de temps annuelle (douze mois) fixée suivant des critères hydrauliques.

3.2.2 Année moyenne Année (fictive) dont les grandeurs hydrauliques constituent les moyennes d'une série cohérente comprenant le plus grand nombre possible d'années. La série annuelle d'après laquelle est calculée l'année moyenne doit être indiquée dans chaque cas.

3.2.3 Année humide Année (fictive) en fonction de critères statistiques, au cours de laquelle les fleuves et rivières ont un débit supérieur à la moyenne.

3.2.4 Année sèche Année (fictive) en fonction de critères statistiques, au cours de laquelle les fleuves et les rivières ont un débit inférieur à la moyenne.

3.2.5 Durée de vidange d'un réservoir Temps minimal nécessaire pour vider un réservoir par les turbines de sa centrale, en supposant absents les apports naturels, depuis le niveau le plus haut jusqu'au niveau le plus bas admis pour son exploitation normale.

3.2.6 Temps de remplissage d'un réservoir Temps qui serait nécessaire pour remplir le réservoir depuis le niveau le plus bas jusqu'au niveau le plus haut admis pour son exploitation normale, si son débit d'alimentation était constant et égal au débit moyen caractéristique corrigé.

3.2.7 Temps de remplissage d'un réservoir à accumulation par pompage Temps nécessaire pour remplir le réservoir supérieur d'un aménagement hydraulique à accumulation par pompage depuis le niveau le plus bas jusqu'au niveau le plus haut admis pour son exploitation normale lorsque les pompes travaillent à plein régime.

Wasserkraftwirtschaft (Hydroenergetik)

Energia Hidroelectrica (energía hidráulica)

3.1 Allgemeine Begriffe

3.1.1 **Wasserkraft (Wasserenergie)** Potentielle Energie (Lageenergie) der Gewässer.

3.1.2 **Wasserkraftwerk** Eine Anlage zur Umwandlung der potentiellen Energie des Wassers in elektrische Energie.

3.1.3 **Laufkraftwerk** Eine Wasserkraftanlage für die laufende (im wesentlichen ungeregelte) Verarbeitung des zufliessenden Wassers.

3.1.4 **Speicherkraftwerk** Ein Wasserkraftwerk, dessen Zufluss mit Hilfe eines Speichers geregelt werden kann.

3.1.5 **Pumpspeicherwerk** Ein Speicherkraftwerk, dessen Speicher ganz oder teilweise durch Pumpen gefüllt wird.

3.1.6 **Gezeitenkraftwerk** Eine Wasserkraftanlage, welche die Fallhöhe ausnützt, die sich infolge der durch Ebbe und Flut bedingten Wasserspiegelschwankungen des Meeres zwischen dem Meer und einem von ihm abgetrennten Becken ergibt.

3.2 Zeitbegriffe

3.2.1 **Wasserwirtschaftsjahr** Eine zwölfmonatige, nach wasserwirtschaftlichen Gesichtspunkten festgesetzte Zeitspanne.

3.2.2 **Regeljahr (Mitteljahr)** Ein fiktives Jahr, dessen wasserwritschaftliche Grössen Mittelwerte einer zusammenhägenden Reihe von möglichst vielen Jahren sind. Die dem Regeljahr (Mitteljahr) zugrundeliegende Jahresreihe ist jeweils anzugeben.

3.2.3 **Nassjahr** Ein nach statistischen Kriterien festgelegtes (fiktives) Jahr mit überdurchschnittlicher Wasserführung.

3.2.4 **Trockenjahr** Ein nach statistischen Kriterien festgelegtes (fiktives) Jahr mit unterdurchschnittlicher Wasserführung.

3.2.5 **Abarbeitungszeit eines Speichers** Die Zeit die zum Abarbeiten der Speicherwassermenge ohne natürliche Zuflüsse bei voller Turbinenleistung (Engpassleistung) benötigt wird.

3.2.6 **Füllzeit eines Speicherbeckens** Die Zeitspanne, die zum Füllen eines Speicherbeckens vom niedrigsten bis zum höchsten betrieblich zulässigen Wasserstand bei mittleren Zuflüssen benötigt wird.

3.2.7 **Füllzeit eines Pumpspeicherbeckens** Die Zeit, spanne, die zum Füllen des Oberbeckens vom niedrigsten bis zum höchsten betrieblich zulässigen Wasserstand eines Pumpspeicherwerkes bei voller Pumpenleistung (ohne Zuflüsse) benötigt wird.

3.1 Conceptos Generales

3.1.1 **Energía hidráulica** Energía potencial de las aguas

3.1.2 **Central hidro-eléctrica** Instalación en la que, la energía potencial de gravedad del agua, es transformada en energía eléctrica.

3.1.3 **Central hidroeléctrica de agua fluyente** "Central hidroeléctrica que no posee embalse regulador".

3.1.4 **Central hidroeléctrica con embalse** "Central hidroeléctrica provista de un embalse que permite regular el caudal de las turbinas".

3.1.5 **Aprovechamiento hidráulico de acumulación por bombeo** Central con embalse, que se llena total o parcialmente por medio de bombas.

3.1.6 **Central maremotriz** Central hidráulica que utiliza la altura de salto entre el mar y un embalse del que está separado, utilizando las variaciones del nivel del mar provocado por las mareas.

3.2 Conceptos Relativos a la División del Tiempo

3.2.1 **Año hidráulico** Período de un año (doce meses) fijado según criterios de economía hidráulica.

3.2.2 **Año medio** Año (ficticio) cuyas características hidráulicas son medias de una serie coherente del mayor número posible de años consecutivos. La serie de años consecutivos en que se basa el año medio o normal ha de ser especificado en cada caso.

3.2.3 **Año húmedo** Año (ficticio), basado en criterios estadísticos, con aportación de agua superior a la media.

3.2.4 **Año seco** Año (ficticio) basado en criterios estadísticos, con aportación de agua inferior a la media anual.

3.2.5 **Tiempo de vaciado de un embalse** Tiempo mínimo necesario para vaciar un embalse a través de las turbinas de su central, suponiendo que no existen aportaciones naturales, desde el nivel máximo al mínimo admitidos en una explotación normal.

3.2.6 **Tiempo de llenado de un embalse** Tiempo que sería necesario para llenar un embalse desde el nivel más bajo al más alto admitidos en explotación normal, si su caudal de alimentación fuese constante e igual al caudal medio característico corregido.

3.2.7 **Tiempo de llenado de un embalse en acumulación por bombeo** Tiempo necesario para llenar el embalse superior vacío de un aprovechamiento hidráulico de acumulación por bombeo, desde el nivel más bajo hasta el más alto admitidos para su explotación normal, trabajando las bombas a plena capacidad.

3

3.3 Terms relating to Location and Head

3.3.1 Catchment area; drainage basin; drainage area The horizontal projection of the area that is in effect drained down to a specified point (e.g. the location of the dam, barrage, weir, etc.).

3.3.2 Dam site The location of the retaining works (dam, barrage, weir or power plant) of a reservoir, constructed so as to create a head of water.

3.3.3 End of backwater; end of upstream reach; end of reservoir pool The transition point at which the watercourse starts to become impounded.

3.3.4 Length of backwater; length of upstream reach; length of reservoir pool The distance measured along the axis of the watercourse between the end of the reservoir pool and the dam site.

3.3.5 Backwater curve A curve representing the increasing elevation of the water surface in the reservoir from the dam, barrage, weir, etc., to the end of the reservoir pool, caused by inflow of water.

3.3.6 Headwater Water upstream of the retaining works or power plant.

3.3.7 Tailwater Water downstream of the retaining works or power plant.

3.3.8 Intake; intake point; headrace Any structure on the upstream dam face or within the reservoir, or in the watercourse, for the purpose of directing water into a confined conduit, tunnel or channel for power production.

3.3.9 Return point; outlet; tailrace The point at which the water after serving to generate electricity is returned to the watercourse.

3.3.10 Scouring reach; deepening reach A tailwater reach affected by deepening or scouring.

3.3.11 Development reach; waterways A stretch of water between the end of the backwater or reservoir pool and the end of the deepening or scouring reach or, where there is no deepening or scouring, the tailrace.

3.3.12 Headwater level The level of the headwater as determined at a reference point on its surface.

3.3.13 Tailwater level The level of the tailwater as determined at a reference point on its surface.

3.3.14 Capacity level; normal reservoir water level; full supply level; full reservoir level The highest level normally allowable in the working of a reservoir or other impounding works, measured at a specified point. It makes no allowance for exceptional excess due to floods.

3.3.15 Lowest operating level; drawdown level The lowest level normally allowable in the working of a reservoir or other impounding works, measured at a specified point.

3.3.16 Gross head; total static head (Aust.); total head difference (UK); critical head (USA); design head (USA) The difference in the level of water destined for the operation of the hydro-electric station between maximum headrace level (or water intake level if there is no headrace) and the final tailrace level.

3.3.17 Net head; rated head (USA) The head actually used by the turbines, i.e. the difference between the level corresponding to manometric height at turbine inlet and, depending on the type of turbine, the tailrace level or average jet level, with adjustment for velocity head.

3.3 Termes Relatifs au Lieux et Hauteurs

3.3.1 Bassin versant/bassin de réception Territoire mesuré en projection horizontale d'où provient effectivement l'eau d'un cours d'eau jusqu'au point considéré.

3.3.2 Lieu de la retenue Point d'implantation de l'ouvrage de retenue d'un barrage destiné à créer une hauteur de chute.

3.3.3 Racine de la retenue Point de raccordement au cours d'eau barré.

3.3.4 Longueur de la retenue Distance entre la racine et le lieu de la retenue, mesurée dans l'axe du cours d'eau.

3.3.5 Ligne de la retenue Ligne représentant le niveau de la surface de l'eau dans le réservoir depuis le lieu de la retenue jusqu'à la racine de la retenue.

3.3.6 Eau d'amont Eau se trouvant en amont d'un lieu de retenue.

3.3.7 Eau d'aval Eau en aval d'un lieu de retenue ou après utilisation de l'énergie hydraulique.

3.3.8 Point de prélèvement, de destockage Lieu (situé sur la face amont du barrage ou dans le réservoir) où l'eau servant à produire de l'énergie est prélevée pour la diriger dans une conduite forcée, un tunnel ou un canal.

3.3.9 Point de restitution Point où l'eau ayant servi à produire de l'énergie est rejetée dans un cours d'eau naturel.

3.3.10 Zone d'approfondissement ou zone d'érosion Zone de l'eau d'aval influencée par un phénomène d'approfondissement ou d'érosion.

3.3.11 Zone d'aménagement Tronçon de cours d'eau situé entre la racine de la retenue et la fin de la zone d'érosion ou - dans l'hypothèse où il n'y aurait pas d'érosion - le lieu de restitution.

3.3.12 Plan d'eau d'amont Niveau du plan d'eau d'amont avec indication du point de mesure.

3.3.13 Plan d'eau aval Niveau du plan d'eau d'aval avec indication du point de mesure.

3.3.14 Niveau le plus haut admis pour l'exploitation d'un réservoir Niveau le plus haut admis normalement d'un réservoir, mesuré en un endroit déterminé (ne tient pas compte des surélévations exceptionnelles dues au passage des crues).

3.3.15 Niveau le plus bas admis pour l'exploitation d'un réservoir Niveau le plus bas admissible pour l'exploitation d'un réservoir, mesuré en un endroit déterminé.

3.3.16 Chute brute d'un aménagement Différence de niveau affectée au fonctionnement de cet aménagement, depuis l'extrémité du remous de la restitution au cours d'eau

3.3.17 Chute nette d'un aménagement Hauteur de chute réellement utilisée par les turbines de l'aménagement, c'est-à-dire différence entre le niveau correspondant à la hauteur manométrique à l'entrée des turbines, compte tenu de l'équivalent en hauteur d'eau de la vitesse de l'eau en ce point et:
— lorsqu'il s'agit de turbines à réaction, le niveau de la restitution majoré de l'équivalent en hauteur d'eau de la vitesse de l'eau à ce point,
— lorsqu'il s'agit de turbines à injecteurs, le niveau moyen des injections.

3.3 Orts- und Höhenbegriffe

3.3.1 Einzugsgebiet Ein in der Horizontalprojektion gemessenes Gebiet, dem der Abfluss an einer bestimmten Stelle eines Gewässers tatsächlich entstammt.

3.3.2 Staustelle, Sperrenstelle, Fallstufe Der Standort eines Absperrbauwerkes (Talsperre, Wehr, Krafthaus) zur Schaffung einer Fallhöhe.

3.3.3 Stauwurzel, Staugrenze Eine Uebergangstelle vom ungestauten zum gestauten Wasserlauf.

3.3.4 Staulänge Eine in der Gewässerachse gemessene Entfernung zwischen Stauwurzel und Staustelle.

3.3.5 Staulinie Ein Wasserspiegel im Längsschnitt der Staulänge.

3.3.6 Oberwasser Das Wasser oberhalb einer Staustelle bzw. vor der Wasserkraftnutzung.

3.3.7 Unterwasser Das Wasser unterhalb einer Staustelle bzw. nach der Wasserkraftnutzung.

3.3.8 Entnahmestelle Der Ort, an dem das der Energiegewinnung dienende Wasser aus dem Wasserlauf entnommen wird.

3.3.9 Rückgabestelle Der Ort, an dem das der Energiegewinnung dienende Wasser wieder in ein natürliches Gewässer eingeleitet wird.

3.3.10 Eintiefungsstrecke Eine durch Eintiefung beeinflusste Strecke des Unterwassers.

3.3.11 Ausbaustrecke Eine Gewässerstrecke zwischen Stauwurzel und Eintiefungsende oder - falls keine Eintiefung vorhanden - Rückgabestelle.

3.3.12 Oberwasserspiegel Die Höhenlage des Oberwassers mit Angabe der Messstelle.

3.3.13 Unterwasserspiegel Die Höhenlage des Unterwassers mit Angabe der Messstelle.

3.3.14 Stauziel Ein festgelegter Normalstau an einer bestimmten Stelle des Stauraumes (ohne Hochwasserschutzraum).

3.3.15 Absenkziel Ein für den Betrieb eines Kraftwerkes festgelegter tiefstzulässiger Wasserspiegel an einer bestimmten Stelle des Stauraumes.

3.3.16 Rohfallhöhe Einer Ausbaustrecke ist der Höhenunterschied der Wasserspiegel am Anfang und am Ende der Ausbaustrecke.

3.3.17 Nettofallhöhe Die Differenz der Energiehöhen beim Ein- und Austritt der Turbine.

3.3 Conceptos Relativos a Emplazamientos y Desniveles

3.3.1 Cuenca vertiente, cuenca receptora Superficie del terreno, medida en proyección horizontal, de la que procede efectivamente el agua de una corriente de agua hasta el punto considerado.

3.3.2 Emplazamiento de la presa Es el lugar ocupado por las obras de la presa de un embalse, contruídas para crear un depósito de agua.

3.3.3 Cola del embalse Punto de transición donde el curso del agua empieza a ser almacenada.

3.3.4 Longitud del embalse Distancia entre la cola del embalse y el emplazamiento de la presa, medida a lo largo del eje de la superficie del agua.

3.3.5 Curva del embalse Curva que representa el nivel de la superficie del agua en el embalse desde la presa hasta la cola del embalse.

3.3.6 Aguas arriba Es la zona aguas arriba de la presa o central.

3.3.7 Aguas abajo Es la zona aguas abajo de la presa o central que utiliza la energía hidráulica.

3.3.8 Toma; punto de toma Lugar (situado en la cara aguas arriba de la central o en el embalse) donde se recoge el agua que ha deservir para producir energía, a fin de encauzarla por una conducción forzada, un túnel o un canal.

3.3.9 Punto de restitución Punto donde el agua es devuelta al río, curso de agua etc., después de haber servido para producir energía.

3.3.10 Zona de erosión; tramo de profundización La zona de la cuenca afectada por profundización o erosión.

3.3.11 Zona de aprovechamiento; tramo ocupado Zona del río entre la cola del embalse y el final del tramo de profundización o erosión o (donde no haya profundización o erosión) el punto de restitución.

3.3.12 Nivel de aguas arriba Nivel de la superficie del agua, aguas arriba, indicando el punto en que se mide.

3.3.13 Nivel de aguas abajo Nivel de la superficie del agua, aguas abajo, indicando el punto en que se mide.

3.3.14 Nivel máximo de explotación El nivel más alto permitido normalmente en un embalse a medir en un lugar determinado (no tiene en cuenta las sobreelevaciones excepcionales debidas al paso de las avenidas).

3.3.15 Nivel mínimo de explotación El nivel más bajo admisible para la explotación de una - central, medida en un lugar determinado.

3.3.16 Salto bruto Diferencia, en determinadas condiciones de caudal y funcionamiento del aprovechamiento, entre el nivel de la cola del embalse y el nivel del agua en la sección transversal de la corriente en que tiene lugar la restitución.

3.3.17 Salto neto En unas condiciones determinadas de aportaciones y de funcionamiento es la altura de salto realmente utilizada por las turbinas del aprovechamiento, es decir, la diferencia entre el nivel correspondiente a la altura manométrica a la entrada de las turbinas, teniendo en cuenta el equivalente en altura de agua de la velocidad del agua en este punto y ei nivel de la restitución.

3.3.18 Geodetic delivery head; static head; pressure head The difference in levels between upper and lower basins in a pumped storage system.

3.3.19 Manometric delivery head; total head; rated net head The geodetic delivery head plus the additional head that is equivalent to the energy required to overcome friction (friction head) and the energy represented by the velocity of discharge (velocity head) in a pumped storage system.

3.3.20 Lost head; loss of head; equivalent loss of head The reduction in useful energy due to friction in the supply conduits, etc., in a water-power scheme, expressed in terms of the head, equivalent to the amount of energy so lost.

3.4 Volume Terms

3.4.1 Discharge flow (inflow, through-flow); flow The volume of water flowing through a cross section in unit time. The term "through-flow" relates more specifically to closed cross sections.

3.4.2 Useful inflow; effective inflow That part of the available flow which after subtracting compensation water and unavoidable losses, is available for generating electricity.

3.4.3 Nominal discharge; rated discharge; turbine flow; operating-flow (SA) (of turbines) The rate of flow for which the turbine is designed.

3.4.4 Maximum usable flow; plant capacity flow; maximum discharge; maximum throughput; maximum operating flow (SA) The maximum flow which the whole of a hydro-electric plant can utilise in continuous operation.

3.4.5 Nominal delivery/discharge; rated delivery/discharge (of pumps) The rate of flow for which the pump is designed.

3.4.6 Cumulative flow (natural, actual, corrected); total discharge; total outflow; total flow (of a watercourse) The volume of water corresponding to the flow (natural, actual or corrected) which passes a given cross-section in a given period of time.

3.4.7 Hydrograph A diagrammatic presentation of observed data in the sequence of their occurrence in time, in the context of water flow.

3.4.8 Frequency distribution curve; duration curve A hydrograph replotted according to the magnitude of the observed data.

3.4.9 Mass curve; cumulative curve; summation curve A hydrograph replotted to give cumulative values.

3.4.10 Hydraulicity; water availability The ratio of total flow in an observed period to the total mean flow for the corresponding period over a long series of years (mean year).

3.4.11 Energy capability factor For a given period, the result obtained by dividing the energy capability of a hydro region by its mean energy capability, both quantities relating to the same period and to the same plant.

3.5 Output Terms

3.5.1 Economic energy potential The water power that is economically exploitable under given conditions.

3.3.18 -Hauteur géodésique (installation de pompage) Différence de niveau entre le bassin supérieur et le bassin inférieur.

3.3.19 Hauteur manométrique d'une pompe Hauteur géodésique augmentée de la hauteur équivalente aux pertes.

3.3.20 Hauteur équivalente aux pertes Hauteur de compensation des pertes de charge dans un ouvrage d'amenée d'eau.

3.4 Termes Relatifs aux Quantités

3.4.1 Débit, écoulement Quantité d'eau s'écoulant pendant l'unité de temps à travers une section (le terme "écoulement" est essentiellement utilisé pour des sections fermées).

3.4.2 Débit utilisable Partie du débit total qui, après déduction d'un volume d'eau obligatoire prévu dans le cahier des charges et des pertes inévitables, reste disponible pour la production d'énergie.

3.4.3 Débit nominal (turbine) Débit pour lequel la turbine a été prévue

3.4.4 Débit maximal turbinable Débit maximal que la turbine peut absorber en régime continu.

3.4.5 Débit nominal (pompe) Débit pour lequel une pompe a été prévue.

3.4.6 Apports Les apports d'un cours d'eau, pendant un intervalle de temps donné et en une section transversale déterminée de son parcours, sont le volume d'eau correspondant au débit qui traverse cette section pendant cet intervalle de temps.

3.4.7 Courbe chronologique Représentation des valeurs observées dans l'ordre chronologique où elles se présentent.

3.4.8 Courbe de fréquence (distribution) Valeurs observées classées en fonction de leur importance.

3.4.9 Courbe cumulée Courbe intégrale d'une courbe chronologique.

3.4.10 Hydraulicité Rapport entre les apports constatés au cours de la période considérée et les apports moyens constatés au cours de la même période sur une longue série d'années (année moyenne).

3.4.11 Indice de productibilité L'indice de productibilité d'une région hydro-électrique pour un intervalle de temps déterminé, est le quotient de sa productibilité par sa productibilité moyenne, toutes deux relatives à cette période et à un même équipement hydro-électrique de la région.

3.5 Termes relatifs à l'énergie

3.5.1 Potentiel énergétique économique Énergie hydraulique qui, dans des conditions données, poourrait être rentablement aménagée.

3.3.18 Geodätische Förderhöhe Einer Pumpe ist der Niveauunterschied zwischen Ober- und Unterbecken.

3.3.19 Manometrische Förderhöhe Einer Pumpe ist gleich der geodätischen Förderhöhe vermehrt um die Verlusthöhe.

3.3.20 Verlusthöhe Der Energiehöhenverlust in einer Wasserführungsstrecke.

3.4 Mengenbegriffe

3.4.1 Abfluss (Zufluss, Durchfluss) Auch Wasserstrom oder Förderstrom, ist die in der Zeiteinheit durch einen Querschnitt fliessende Wassermenge. (Der Begriff "Durchfluss" wird vorwiegend für geschlossene Querschnitte verwendet).

3.4.2 Nutzbarer Zufluss Jener Teil des Gesamtzuflusses, der nach Abzug der Pflichtwassermenge und unvermeidbarer Verluste für die Energiegewinnung verfügbar ist.

3.4.3 Nenndurchfluss (Turbine) Jener Durchfluss, für den die Turbine ausgelegt ist.

3.4.4 Schluckfähigkeit (max. Durchsatz) Der höchstmöglicher Durchfluss durch die Turbine bei Dauerbetrieb.

3.4.5 Nennförderstrom (Pumpe) Der Förderstrom, für den eine Pumpe ausgelegt ist.

3.4.6 Abfluss-Summe Ein für eine bestimmte Zeitspanne summierter (integrierter) Abfluss.

3.4.7 Ganglinie Eine zeichnerische Darstellung von Beobachtungswerten in der Reihenfolge ihres zeitlichen Auftretens.

3.4.8 Dauerlinie Eine nach der Grösse der Beobachtungswerte geordnete Ganglinie.

3.4.9 Summenlinie Eine Integralkurve einer Ganglinie.

3.4.10 Hydraulizität Das Verhältnis zwischen dem Gesamtabfluss im betrachteten Zeitraum und dem entsprechenden mittleren Gesamtabflussim Regeljahr.

3.4.11 Erzeugungskoeffizient Das Verhältnis zwischen der Erzeugung im betrachteten Zeitraum und der entsprechenden mittleren möglichen Erzeugung im Regeljahr.

3.5 Arbeitsbegriffe

3.5.1 Oekonomisches energetisches Potential Die Wasserkraft, die unter gegebenen Umständen wirtschaftlich ausbaufähig ist.

3.3.18 Altura geodésica (instalación de bombeo Diferencia de nivel entre el embalse superior y el embalse inferior.

3.3.19 Altura manométrica de una bomba Es la altura geodésica más la altura equivalente a las pérdidas de carga.

3.3.20 Pérdida de carga Altura de compensación de las pérdidas de carga en un aprovechamiento hidráulico.

3.4 Conceptos Cuantitativos

3.4.1 Caudal; escorrentía Volumen de agua fluyendo a través de una sección en la unidad de tiempo.

3.4.2 Caudal utilizable Parte del caudal total que queda disponible para la producción de energía, una vez deducidos el caudal no utilizable según las condiciones de la concesión y las pérdidas inevitables.

3.4.3 Caudal nominal (turbina) Caudal para el que está prevista la turbina.

3.4.4 Caudal máximo turbinable Máximo caudal utilizable por una turbina, en régimen continuo.

3.4.5 Caudal nominal (bomba) Caudal para el que está prevista una bomba.

3.4.6 Aportaciones De una corriente de agua, durante un intervalo de tiempo dado y en una sección transversal determinada de su recorrido, son los volúmenes de agua que corresponden al caudal que atraviesa dicha sección durante el espacio de tiempo de referencia.

3.4.7 Curva cronológica Representación gráfica de los datos observados, en el mismo orden cronológico en que se presentan.

3.4.8 Curva de caudales clasificados (curva de frecuencia) Representación gráfica de los valores observados clasificados en orden de su magnitud.

3.4.9 Curva de valores acumulados Curva integral de una curva cronológica.

3.4.10. Hidraulicidad Relación entre las aportaciones dentro del período observado y las aportaciones medias correspondientes a un mismo período a lo largo de una serie de años (año medio).

3.4.11 Coeficiente de utilización De una región hidroeléctrica durante un determinado intervalo de tiempo, es la relación entre la producción en un periodo de tiempo determinado y la posible producción en el mismo período de tiempo correspondiente a un ano medio, referidos ambos ano ese período y a un mismo equipamiento hidroeléctrico de la región.

3.5 Conceptos Relativos a la Energía

3.5.1 Potencial económicamente explotable Energía hidráulica que puede aprovecharse rentablemente en determinadas condiciones.

3.5.2 **Energy capability of a hydro-electric power station** The maximum amount of electricity which the cumulative corrected flow, recorded during a given period, would allow it to produce under the best conditions during that period.

3.5.3 **Energy capability of a reservoir; equivalent electrical energy capacity of a reservoir** (Aust.) The amount of electricity which could be produced from its own generating station and generally from all stations downstream thereof by the complete drawing off of its "useful water capacity", such drawing off being assumed to be carried out without natural inflows at a rate excluding all water losses.

3.5.4 **Energy capability of a pumped storage station during turbine operation** The electrical energy that could be generated by the turbines when their storage reservoirs are initially full.

3.5.5 **Energy absorbed by storage pumping; energy absorbed by pumping** The energy consumed by the motor pump in raising the water into the upper reservoir for the generation of electricity. It should include the energy consumed by the auxiliary equipment and losses during pumping.

3.5.6 **Conversion efficiency of a pumped storage cycle; pumped storage index** The ratio of the electrical energy derived from pumping to the electrical energy absorbed by pumping relative to the same quantity of water pumped, during one cycle.

3.6 Storage Terms

3.6.1 **Daily storage** Storage in which the reservoir has a daily filling and emptying cycle.

3.6.2 **Weekly storage** Storage in which the reservoir has a weekly filling and emptying cycle.

3.6.3 **Seasonal storage** Storage in which the reservoir has a seasonal filling and emptying cycle.

3.6.4 **Annual storage** Storage in which the reservoir has an annual filling and emptying cycle.

3.6.5 **Storage of more than one year** Storage in which the reservoir is able to even out fluctuations in water availability over a period in excess of one year.

3.6.6 **Useful water capacity; working storage** (Aust.) The volume of water which a reservoir can hold between the lowest and highest levels normally allowable in use.

3.6.7 **Flood control storage basin; flood water storage volume** (Aust.); **flood water retention area** (SA) The storage volume of a reservoir that exists between the highest level normally allowable in its working and the maximum possible water level; alternatively the area over which such maximum possible water level would extend.

3.5.2 **Productibilité d'un aménagement hydro-électrique** La productibilité d'un aménagement hydro-électrique pendant un intervalle de temps déterminé, est la quantité maximale d'énergie électrique que l'ensemble des apports corrigés constatés pendant l'intervalle de temps considéré lui permettrait de produire dans les conditions les plus favorables.

3.5.3 **Capacité en énergie électrique d'un réservoir** Quantité d'énergie électrique qui serait produite dans sa propre centrale et, en général, dans toutes les centrales situées à l'aval de celle-ci par la vidange complète de sa "capacité utile en eau", cette vidange étant supposée faite en l'absence d'apports naturels et à une cadence excluant toute perte d'eau.

3.5.4 **Capacité en énergie électrique d'une centrale à accumulation par pompage pendant le fonctionnement des turbines** Énergie électrique susceptible d'être produite par les turbines, le réservoir supérieur étant plein initialement.

3.5.5 **Energie absorbée pour le pompage dans le cas d'une centrale à accumulation pendant le fonctionnement des pompes** Énergie électrique consommée par les groupes moto-pompes pour l'élévation de l'eau dans les réservoirs en vue de la production d'énergie, y compris l'énergie consommée par les auxiliaires et les pertes.

3.5.6 **Rendement du cycle d'un réservoir de centrale à accumulation par pompage** Le rendement du pompage pendant un cycle est le rapport entre l'énergie électrique produite à partir de pompage et l'énergie électrique absorbée pour le pompage relatives à la même quantité d'eau pompée.

3.6 Stockage

3.6.1 **Réservoir journalier** Réservoir ayant un cycle journalier pour le remplissage et la vidange.

3.6.2 **Réservoir hebdomadaire** Réservoir ayant un cycle hebdomadaire pour le remplissage et la vidange.

3.6.3 **Réservoir saisonnier** Réservoir ayant un cycle saisonnier pour le remplissage et la vidange.

3.6.4 **Réservoir annuel** Réservoir ayant un cycle annuel pour le remplissage et la vidange.

3.6.5 **Réservoir pluri-annuel** Réservoir qui permet de compenser au-delà d'une année les variations de l'hydraulicité.

3.6.6 **Capacité utile en eau** Volume d'eau d'un réservoir compris entre le niveau le plus haut et le niveau le plus bas admis pour son exploitation normale.

3.6.7 **Zone inondable** Volume d'un réservoir compris entre le niveau le plus haut admis pour son exploitation normale et le niveau d'eau maximal possible.

3.5.2 Arbeitsvermögen eines Wasserkraftwerkes Die innerhalb eines bestimmten Zeitabschnittes aus dem nutzbaren Zufluss erzeugbare elektrische Arbeit.

3.5.3 Arbeitsvermögen eines Speichers Die Elektroenergie, die in einem gegebenen Wasserkraftwerk durch die Nutzung der Wassermenge, die dem Nutzinhalt des gegebenen Speicherbeckens entspricht, erzeugt werden kann. Das Arbeitsvermögen unterliegender Wasserkraftwerke wird im allgemeinen miteinbezogen.

3.5.4 Arbeitsvermögen eines Pumpspeicherwerkes bei Turbinenbetrieb Die mit dem vollen Oberbecken erzeugbare elektrische Arbeit.

3.5.5 Pumpenenergieaufwand eines Pumpspeicherwerkes bei Pumpbetrieb Die elektrische Arbeit, die zur Förderung des Speicherwassers aufgewendet werden muss, einschliesslich Eigenverbrauch und Verluste.

3.5.6 Wirkungsgrad eines Pumpspeicherzyklus Das Verhältnis der mit dem Nutzinhalt des Oberbeckens erzeugten elektrischen Arbeit zur elektrischen Arbeit, die für die Förderung dieses Nutzinhaltes aufgewendet werden muss.

3.6 Speicherung

3.6.1 Tagesspeicher Ein Speicher mit ausgeprägtem Tageszyklus des Aufstauens und Entleerens.

3.6.2 Wochenspeicher Ein Speicher mit ausgeprägtem Wochenzyklus des Aufstauens und Entleerens.

3.6.3 Saisonspeicher Ein Speicher mit ausgeprägtem Saisonzyklus des Aufstauens und Entleerens.

3.6.4 Jahresspeicher Ein Speicher mit ausgeprägtem Jahreszyklus des Aufstauens und Entleerens.

3.6.5 Ueberjahresspeicher Ein Speicher, der die Schwankungen des Waserdargebotes über ein Jahr hinaus auszugleichen gestattet.

3.6.6 Nutzinhalt (Nenninhalt) Der Zwischen Stauziel und Absenkziel befindliche Speicherraum.

3.6.7 Hochwasserschutzraum Der über dem Stauziel bis zum höchstmöglichen Wasserspiegel befindliche Speicherraum.

3.5.2 Producibilidad de un aprovechamiento hidroeléctrico En un intervalo de tiempo determinado es la máxima cantidad de energía que le permitirían producir, en las condiciones más favorables, el conjunto de aportciones corregidas correspondientes al intervalo de tiempo considerado.

3.5.3 Capacidad en energía eléctrica de un embalse Cantidad de energía eléctrica que se produciría en su propia central y, en general, en todas las situadas aguas abajo de ésta con el vaciado completo de su "capacidad útil en agua" en el supuesto de que este vaciado se haga en auscenia de aportaciones naturales y a un ritmo que excluya cualquier pérdida de agua.

3.5.4 Capacidad de producción de una central de acumulacion por bombeo funcionando con turbinas Energía eléctrica que puede producirse por turbinas con el embalse superior inicialmente lleno.

3.5.5 Energía consumida por el bombeo en caso de una central de acumulación por bombeo Energía eléctrica absorbida por los grupos moto-bombas para la elevación del agua a los embalses con vistas a la producción de energía, incluida la energía consumida en los servicios auxiliares y las pérdidas de transformación durante el bombeo.

3.5.6 Rendimiento del ciclo del embalse de una central de acumulación por bombeo Es la relación de la energía eléctrica producida por el agua que ha sido acumulada por bombeo en el embalse superior, a la energía eléctrica absorbida por el bombeo en la elevación de la misma cantidad de agua.

3.6 Almacenamiento

3.6.1 Embalse diario Embalse cuyo ciclo de llenado y vaciado es de un día.

3.6.2 Embalse semanal Embalse cuyo ciclo de llenado y vaciado es una semana.

3.6.3 Embalse estacional Embalse cuyo ciclo de llenado y vaciado dura una estación.

3.6.4 Embalse anual Embalse cuyo ciclo de llenado y vaciado dura un año.

3.6.5 Embalse hiperanual Embalse que permite compensar las variaciones de hidraulicidad en ciclos de más de un año de duración.

3.6.6 Capacidad útil en agua La capacidad útil en agua de un embalse es el volumen de agua que puede contener entre los niveles mínimo y máximo admitidos en su explotación normal.

3.6.7 Zona inundable Zona de un embalse comprendido entre el nivel más alto admitido para su explotación normal y el máximo nivel de agua posible.

3

Section 4

Mining and Processing of Solid Fuels
Extraction et préparation des combustibles solides
Gewinnung und Verarbeitung fester Brennstoffe
Extraccion y Preparacion de los Combustibles Solidos

Mining and Processing of Solid Fuels

4.1 Classification of Solid Fuels

4.1.1 Run of mine fuel; fuel as mined; raw fuel The fuel immediately after it has been mined, before any subsequent processing.

Note More usually the particular type of fuel under consideration would be stated, e.g. *run-of-mine coal, raw lignite*.

4.1.2 Hard coal Combustible, solid, black, fossil carbonaceous sedimentary deposit with a gross calorific value over 24 MJ/kg (approx 5700 kcal/kg; 10,260 Btu/lb) on the moist ash-free basis, being the gross calorific value of the coal when in equilibrium with air at 30°C and 96% relative humidity, calculated on the ash-free basis. See second note below term 4.1.3.

Note In view of the difficulty of differentiating in borderline cases between hard coal and brown coal/lignite, defined in 4.1.3, the following further identification reactions may be applied:

Streak colour: black.

Humic acid reaction with KOH: colourless, wine yellow or greenish, not reddish.

Lignin reaction with HNO_3: none.

4.1.3 Brown coal; lignite Combustible, solid, black to brown, fossil carbonaceous sedimentary deposit. Until reliable parameters for differentiation of brown and hard coals are worked out and confirmed, coals considered in each country as brown, on the basis of a number of other characteristics, should be classified as brown coals regardless of their calorific value, i.e. including the cases when the gross calorific value of the coal in equilibrium with air at 30°C and 96% relative humidity is more than 24 MJ/kg on the ash-free basis. See second note below.

Note 1 In view of the difficulty of differentiating in borderline cases between hard coal, defined in 4.1.2, the following further identification reactions may be applied:

Streak colour: light to dark brown.

Humic acid reaction with KOH: brown.

Lignin reaction with HNO_3: orange to reddish.

Note 2 Conventional use of the terms "hard coal", "brown coal" and "lignite" varies from country to country. Taking the naturally occurring solid fuels by rank, namely, wood, peat, lignite/brown coal, sub-bituminous coal, bituminous coal, semi-bituminous coal, semi-anthracite and anthracite, in the USA and Canada, anthracite is considered to be hard coal, bituminous coal is considered to be soft coal, and sub-bituminous coal is synonymous with black lignite; under the above-mentioned international classification the term hard coal includes both anthracite and bituminous coal.

4.1.4 Peat Combustible, soft, porous or compressed, fossil sedimentary deposit of plant origin with high water content (up to 90% in the raw state), easily cut, of light to dark brown colour.

4.1.5 Wood Requires no definition. See second note to 4.1.3 above.

Extraction et préparation des combustibles solides

4.1 Classification des Combustibles

4.1.1 Combustible brut Combustible considéré immédiatement après son extraction, avant tout traitement ultérieur.

4.1.2 Charbon Sédiment fossile organique solide, combustible, noir, dont le pouvoir calorifique superieur est supérieur à 24 MJ/kg ($\cong$ 5700 kcal/kg), en considérant la substance sans cendres et dont la teneur en eau est celle qui s'établit à une température de 30°C et pour une humitidé relative de l'air de 96%.
(International Classification of Hard Coal by Type : E/ECE/247. United Nations Publication, Genève, Août 1956).
Remarque : en raison des difficultés concernant la délimitation entre charbons et lignites indiquée en 4.1.3., les autres réactions de détermination suivantes sont applicables : Trace sur une feuille de papier : noire
Réaction à l'acide humique avec KOH: incolore, jaune vin ou verdâtre, non rougeâtre.
Réaction de la lignine avec HNO_3 : aucune coloration.

4.1.3 Lignite Sédiment fossile organique, combustible, brun à noir dont le p.c.s. est inférieur à 24 MJ/kg (5700 kcal/kg), en considérant la substance sans cendres et dont la teneur en eau est celle qui s'établit à une température de 30°C et pour une humitité relative de l'air de 96%.
(Classification internationale — ISO/DIS 2950-72)
Remarque : en raison des difficultés de la délimitation entre charbons et lignites, les autres réactions de détermination suivantes sont applicables :
Trace sur une feuille de papier : brun clair à brun foncé. Réaction à l'acide humique avec KOH: coloration brune. Réaction à la lignine avec HNO_3 : coloration orangeé à rougeâtre.

4.1.4 Tourbe Sédiment fossile d'origine végétale, poreux ou comprimé combustible à haute teneur en eau (jusqu'à environ 90% sur brut) facilement rayé, de couleur brun clair à brun foncé.

4.1.5 Bois Sans définition.

Gewinnung und Verarbeitung fester Brennstoffe

4.1 Einteilung der Brennstoffe

4.1.1 Rohbrennstoff Brennstoff unmittelbar nach seiner Gewinnung, vor seiner weiteren Verarbeitung.

4.1.2 Steinkohle Festes, brennbares, schwarzes, fossiles organisches Sediment, dessen Verbrenngswärme mehr als 24 MJ/kg ($\cong$ 5'700 kcal/kg) beträgt, bezogen auf aschefreie Substanz mit einem Wassegehalt, der sich bei einer Temperatur von 30°C und einer relativen Luftfeuchte von 96% einstellt.
Internationale Klassifikation von Steinkohle (International Classification of Hard Coal by Type : E/ECE/247. United Nations Publication. Genf August 1956).
Anmerkung : Wegen der Schwierigkeiten bei der Abgrenzung zu 4.1.3. "Braunkohle" gelten folgende weiteren Erkennungsreaktionen :
Strichfarbe : schwarz
Huminsäurereaktion mit KOH : farblos, weingelb oder grünlich, nicht rötlich
Ligninreaktion mit HNO_3 : keine

4.1.3 Braunkohle Festes, brennbares, braunes bis schwarzes, fossiles organisches Sediment, dessen Verbrennungswärme weniger als 24 MJ/kg (5'700 kcal/kg) beträgt, bezogen auf aschefreie Substanz mit einem Wassergehalt, der sich bei einer Temperatur von 30°C und einer relativen Luftfeuchte von 96% einstellt.
Internationale Klassifikation (ISO/DIS 2950-72)
Anmerkung : Wegen der Schwierigkeiten bei der Abgrenzung zu "Steinkohle" (4.1.2) gelten folgende weiteren Erkennungsreaktionen :
Strichfarbe : hell- bis dunkelbraun
Huminsäurereaktion mit KOH : braun
Ligninreaktion mit HNO_3 : orange bis rötlich

4.1.4 Torf Weiches, lockeres bis gepresstes, brennbares fossiles Sediment pflanzlichen Ursprungs mit hohem Wassergehalt (bis etwa 90% (i.roh)), stechbar, von hellbrauner bis dunkelbrauner Farbe.

4.1.5 Holz Ohne Definition

Extraccion y Preparacion de los Combustibles Solidos

4.1 Clasificación de los Combustibles

4.1.1 Combustible bruto Combustible inmediatamente después de su extracción, antes de cualquier tratamiento ulterior.

4.1.2 Antracita, carbón de piedra, carbón duro Sedimento fósil orgánico sólido, combustible, negro, cuyo p.c.s. (poder calorífico superior) es mayor de 24 MJ/kg ($\cong$ 5.700 Kcal/kg) referido a la sustancia sin cenizas, cuyo contenido de agua sea el que se establece a una temperatura de 30° C y con una humedad relativa del aire del 96% (Internacional Classification of Hard Coal by Type E/ECE/247, United Nations Publications, - Ginebra. Agosto 1.956).
Observación: Debido a las dificultades relativas a la delimitación entre carbones y lignitos indicada en 4.1.3 son de aplicación las otras reacciones de determinación siguientes:
Traza sobre una hoja de papel: negra
Reacción al ácido húmico con KOH: incoloro, amarillo vinoso o verdoso, no rojizo.
Reacción a la lignina con $H NO_3$: ninguna

4.1.3 Lignito Sedimento fósil orgánico, combustible, pardo a negro cuyo p.c.s. es inferior a 24 MJ/Kg. (5.700 Kcal/kg) referido a la sustancia sin cenizas cuyo contenido de agua sea el que se establece a una temperatura de 30° C y con una humedad relativa del 96%.
(Clasificación internacional - ISO/DIS 2050-72)
Observación: Debido a las dificultades relativas a la delimitación entre carbones y lignitos, son de aplicación las otras reacciones de determinación siguientes:
Traza sobre una hoja de papel: marrón claro o marrón oscuro
Reacción del ácido húmico con KOH: Coloración marrón.
Reacción a la lignina con $H NO_3$: coloración naranja o rojiza.

4.1.4 Turba Sedimento fósil de origen vegetal, poroso o comprimido, combustible, de alto contenido de agua (hasta alrededor del 90% sobre producto bruto) fácilmente rayable, de color marrón claro a marrón oscuro.

4.1.5 Madera No requiere definición

4

4.1.6 **Trash; rubbish; refuse; solid waste** All solid waste material that occurs in households, commercial activities, municipal plants, industry, etc.

4.1.7 **Product of processing** The product derived from subjecting the raw fuel to established processing steps. In normal contexts a more specific term would be used, e.g. "coal products" or "peat products".

4.1.8 **Prepared coal; treated coal; processed coal** (SA) A product obtained from raw coal, which by processes of preparation, such as grading, cleaning, grinding, crushing, dewatering, blending, is made suitable for its specific application.

4.1.9 **Sized coal; graded coal; screened goal** Coal screened within specified size limits.

4.1.10 **Cleaned coal; sorted coal** (SA) Coal that has undergone treatment to reduce its mineral-matter content (ash, sulphur) to within narrow limits.

4.1.11 **Washed coal** The end product of mechanical, wet (or dry) cleaning, rich in dry, ash-free coal.

4.1.12 **Middlings (as final product)** A product of the preparation of coal which, by reason of its ash content, is too poor in quality for ready sale, but contains too much combustible matter to be discarded.

4.1.13 **Washery refuse; washery dirt** That part of the end product of mechanical, dry or wet cleaning that contains such a high percentage of waste material that it is discarded.

4.1.14 **Dried brown coal; dried lignite** Brown coal/lignite whose moisture content has been reduced by drying.

Note Dried brown coal/lignite may be the sum of the product leaving the drier and the dust entrained in the drying medium.

4.1.15 **Briquette; ovoid** A shaped fuel made by compressing pre-treated solid fuel fines in a press with or without a binder. The granular size of the feed material and the briquette or ovoid itself may be varied to suit the eventual application of the fuel.

4.1.16 **Coke** Solid fuel obtained from coal by heating in the absence of air.

4.1.17 **High temperature coke** The solid residue of the distillation of coal at temperatures above 800°C. This lower temperature limit, however, does not apply universally; in some countries (e.g. French and German-speaking countries) the lower limit is 1000°C for hard coal and 900°C for brown coal coke.

4.1.18 **Low temperature coke; semi-coke** The solid residue of the low temperature distillation of coal (500 to 800°C). In the case of brown coal the temperature would be 400 to 600°C; in the case of peat, 350 to 550°C.

4.1.19 **Formed coke** Coke made from briquetted or pelletised coal for metallurgical purposes.

4.1.20 **Solid smokeless fuel** A fuel whose natural properties or whose properties resulting from special treatment are such that when burned the fuel emits only limited quantities of visible solid or liquid substances (e.g. ash, soot, tar) in the flue gases.

4.1.21 **Run-of-mine output; as-mined output; raw fuel output** The quantity of fuel mined before preparation and including inerts, but excluding inerts that do not pass through the preparation stage.

4.1.22 **Saleable output** The quantity of fuel produced after preparation. In statistical returns the fuel should be classified, e.g. by mineral matter content, calorific value.

4.1.6 **Ordures, déchets, résidus** Tous les déchets solides récoltés dans les foyers domestiques, dans les exploitations commerciales, les installations collectives, l'industrie, etc. . .

4.1.7 **Produit préparé** Produit provenant d'une opération déterminée de préparation de combustibles bruts.

4.1.8 **Charbon préparé** Produit obtenu à partir du charbon brut, qui a été rendu propre à l'utilisation concernée par des procédés et des processus de préparation tels que le classement par calibres, le triage, l'épuration, la comminution, l'égouttage, le mélange.

4.1.9 **Charbon classé, charbon calibré** Charbon appartenant à une classe de granulométrie déterminée.

4.1.10 **Charbon trié, charbon épuré** Charbon préparé appartenant à des domaines étroits de teneur en matières minérales (cendres, soufre). (Cf. 4.4.2.)

4.1.11 **Charbon lavé** Produit final enrichi en charbon pur, résultant d'une épuration mécanique, par voie sèche ou humide.

4.1.12 **Mixtes** Produit final mélangé, résultant d'une épuration mécanique, par voie sèche ou humide.

4.1.13 **Déchets** Produit final enrichi en stérile, résultant de l'épuration mécanique, par voie sèche ou humide.

4.1.14 **Lignite séché** Lignite dont la teneur en eau a été abaissée par séchage.

Remarque : le lignite séché peut être l'ensemble du produit sortant du sécheur et du poussier entraîné par les vapeurs.

4.1.15 **Agglomérés, briquettes et boulets** Combustible moulé et obtenu par compression, après préparation préalable d'un combustible de fine granulométrie, éventuellement en mélange avec des liants. La dimension des agglomérés ainsi que leur granulométrie peuvent être variables en fonction de l'utilisation.

4.1.16 **Coke** Combustible solide obtenu à partir du charbon par pyrolyse à l'abri de l'air.

4.1.17 **Coke de haute température, coke** Coke obtenu par cokéfaction de charbons à des températures supérieures à 1000°C ou de lignites à des températures supérieures à 900°C.

4.1.18 **Coke de basse température, semi-coke** Coke obtenu par cokéfaction de charbons à des températures de 600°C environ ou de lignites à des températures de 400 à 600°C, ou, enfin, de tourbe à des températures de 350 à 550°C.

4.1.19 **Coke moulé** Coke fabriqué à partir d'agglomérés de charbon.

4.1.20 **Combustible sans fumée** Combustible possédant des caractéristiques naturelles ou résultant d'un traitement particulier telles que, lors de la combustion, il ne fournit dans les produits de la combustion (fumées) qu'une faible quantité de matières visibles solides et liquides, par exemple : cendres, suies, goudron.

4.1.21 **Extraction brute** Quantité extraite considérée avant la préparation et comprenant les stériles, sauf ceux qui ne passent pas par la préparation.

4.1.22 **Production marchande** Quantité considérée après la préparation. Dans les données statistiques, il y a lieu d'indiquer la base de référence, par exemple : la teneur en internes, le p.c.i.

4.1.6 Müll Sämtliche festen Abfälle, die in Haushaltungen, Gewerbebetrieben, kommunalen Anlagen, in der Industrie usw. anfallen.

4.1.7 Verarbeitungsprodukt Erzeugnis, das aus bestimmten Verarbeitungsverfahren von Rohbrennstoffen hervorgegangen ist.

4.1.8 Aufbereitete Kohle Aus der Rohkohle hergestelltes Erzeugnis, das durch Verfahren und Vorgänge in der Aufbereitung wie Klassieren, Sortieren, Zerkleinern, Aufschliessen, Entwässern, Mischen für den jeweiligen Verwendungszweck geeignet gemacht worden ist.

4.1.9 Klassierte Kohle Kohle mit bestimmten Korngrössenbereichen.

4.1.10 Sortierte Kohle Nach Grösse geordnete und gereinigte Kohle

4.1.11 Waschkohle (gewaschene Kohle) An Reinkohle angereichertes Enderzeugnis der mechanischen, trockenen oder nassen Sortierung.

4.1.12 Mittelgut An Verwachsenem angereichertes Enderzeugnis der mechanischen, trockenen oder nassen Sortierung.

4.1.13 Waschberge An Reinbergen angereichertes Enderzeugnis der mechanischen, trockenen oder nassen Sortierung.

4.1.14 Trockenbraunkohle Braunkohle, deren Wassergehalt durch Trocknen gesenkt wurde.

Anmerkung : Trockenbraunkohle kann die Summe des Trockneraustragsgutes und des Brüdenstaubes sein.

4.1.15 Brikett Geformter Brennstoff, der nach Vorbehandlung feinkörniger Brennstoffe durch Verpressen, gegebenenfalls im Gemenge mit Bindemitteln, hergestellt worden ist. Die Körnung des Brikettiergutes kann je nach Verwendungszweck unterschiedlich sein.

4.1.16 Koks Aus Kohle durch Erhitzen unter Luftabschluss erhaltener fester Brennstoff.

4.1.17 Hochtemperaturkoks Durch Verkokung von Steinkohlen bei Temperaturen über 1000°C oder von Braunkohlen bei Temperaturen über 900°C hergestellter Koks.

4.1.18 Schwelkoks Durch Verkokung von Steinkohlen bei Temperaturen von etwa 600°C oder von Braunkohlen bei Temperaturen von 400 bis 600°C oder von Torf bei Temperaturen von 350 bis 550°C hergestellter Koks.

4.1.19 Formkoks Aus Kohlenbriketts oder -pellets hergestellter Koks.

4.1.20 Raucharmer Brennstoff Brennstoff mit solchen natürlichen oder durch besondere Behandlung eingestellten Eigenschaften, dass er bei der Verbrennung in die Abgase (Rauchgase) nur wenig sichtbare feste und flüssige Stoffe, z.B. Asche, Russ, Teer abgibt.

4.1.21 Rohfördermenge Geförderte Menge vor der Aufbereitung einschliesslich Berge, jedoch ohne die Berge, die nicht durch die Aufbereitung gehen.

4.1.22 Verkaufsfähige Fördermenge Geförderte Menge nach der Aufbereitung. Bei statistischen Angaben ist die Bezugsbasis, z.B. Ballastgehalt, Heizwert zu nennen.

4.1.6 Desperdicios, desechos, residuos (plural) Residuos sólidos que se producen en los hogares domésticos, actividades comerciales, instalaciones colectivas, industria, etc.

4.1.7 Producto preparado Producto obtenido al tratar los combustibles brutos por procesos determinados.

4.1.8 Carbón preparado Producto obtenido a partir del carbón bruto, convertido en utilizable al haber sido sometido a procedimientos y procesos de preparación, tales como clasificación, cribado, limpieza, machaqueo, secado, mezcla.

4.1.9 Carbón clasificado por tamaños Carbón perteneciente a una determinada granulometría.

4.1.10 Carbón paro, carbón escogido Carbón preparado conteniendo proporciones minimas de impurezas (cenizas, azufre) (Ver 4.4.2).

4.1.11 Carbón lavado Producto final de un proceso de lavado mecánico, por via seca o húmeda, rico en carbón puro.

4.1.12 Mixtos (plural) Producto final, mezclado, resultante de un lavado mecánico, por vía seca o húmeda.

4.1.13 Estériles del lavado Producto final del lavado mecánico, por vía seca o húmeda, que contiene alta proporción de materia estéril.

4.1.14 Lignito secado Lignito cuyo contenido en agua se reduce mediante secado.

Observación: El lignito secado puede ser el conjunto del producto que sale del secado y del polvo arrastrado por los vapores.

4.1.15 Aglomerados, briquetas Es un combustible molido fabricado por compresión en una prensa, de finos de combustible sólido con adición o sin elia de un material aglomerante. El tamaño del aglomerado así como su granulometría puede ser distinto según el uso a que se destine.

4.1.16 Coque Combustible sólido obtenido por pirólisis del carbón en ausencia del aire.

4.1.17 Coque de alta temperatura Coque obtenido por coquización de carbones a temperaturas superiores a 1.000° C o lignitos a temperaturas superiores a 900° C.

4.1.18 Coque de baja temperatura, semi-coque Coque obtenido por coquización de carbones a temperaturas de alrededor de 600°C o de lignitos a temperaturas entre 400 y 600°C o, por último, de turba a temperaturas entre 350 y 550° C.

4.1.19 Coque pre-formado Coque obtenido a partir de aglomerados de carbón.

4.1.20 Combustible sin humo Combustible que posee características naturales o resultado de un tratamiento particular tal que, los humos producto de su combustión no contienen más que cantidades muy pequeñas de materias visibles sólidas y líquidas, por ejemplo, cenizas, hollín, a lquitrán.

4.1.21 Producción bruta Cantidad extraída considerada antes de la preparación y que incluye los estériles, salvo aquellos que no han de pasar por la preparación.

4.1.22 Producción vendible Cantidad considerada después de la preparación. En los datos estadísticos debe indicarse sus características, por ejemplo, su contenido en material inerte, su p.c.i. (poder calorífico inferior).

4

4.1.23 Ton/tonne of coal equivalent (tce)/(tece); standard coal equivalent (SA) A common unit employed to enable oil, gas, nuclear and hydro-electricity, hard coal, brown coal and other forms of energy to be directly compared from the point of view of their fuel value; conversion of fuels to their coal equivalents depends mainly, but not necessarily exclusively, on their calorific value. The actual conversion factors used may vary from one country or authority to another. In French and German speaking countries 29.3 GJ (approx. 7000 Mcal) is the net calorific value of one tonne of coal equivalent for the purposes of calculation (i.e. 282 therms per ton of coal equivalent).

4.1.23 Tonne d'équivalent charbon ou t.e.c. Combustible d'un p.c.i. de 29,3 MJ/kg ($\cong$7000 kcal/kg) considéré comme référence quantitative.

4.2 Solid Fuel Deposits—Under Consideration

4.2 Gisements—A l'Etude

4.3 Mining (and Transport)

4.3.1 Deep mining Mining in which access to the mineral deposits is obtained by means of shafts and underground workings.

4.3.2 (Exploratory) development work Mining works whose purpose is to locate the presence of economic deposits of coal and to establish their extent and initial access to them.

4.3.3 (Preparatory) development work Mining works whose purpose is to open up proven coal deposits, by planning the layout of the mine and establishing access to the coal deposit in preparation for the actual extraction.

4.3.4 Mining; winning; extraction The extraction of the contents of the seam from the geological formation in which it is located.

4.3.5 Transport; transportation; haulage; conveying Collective terms for the forward transport from the point of extraction of the substance mined. In statistical returns the limits of the zone of operations should be entered.

4.3.6 Support Equipment for ensuring roadways, galleries, etc., in mines against collapse.

4.3.7 Mine pumping; drainage; dewatering All procedures or plants and installations above or below ground for keeping water out of the mine workings, as well as for collecting, clarifying and carrying off incoming water.

4.3.8 Ventilation All procedures and installations for providing fresh air to mine workings, diluting and removing toxic and vitiated air and firedamp, and improving the mine climate.

4.3.9 Production days Working days in which coal, etc. is actually produced.

4.3.10 Opencast mining; open-cut mining; open-pit mining (Can.); strip mining (USA) The mining of deposits of one or more seams thickness from the surface after removal of the overburden.

4.3.11 Deep opencast etc. mining This term is the same as 4.3.10 above; the addition of the adjective "deep" has in English no specific meaning other than the normal English usage of the word and is unlike the specific German meaning in this context, i.e. opencast mining at a depth of more than 200 metres.

4.3 Extraction

4.3.1 Exploitation souterraine Exploitation de gisements minéraux utiles, qui sont reconnus par galeries ou puits et exploités à partir d'excavations minières.

4.3.2 Travaux préparatoires au rocher Terme générique désignant la réalisation de toutes excavations minières qui ont pour objet la recherche et la découverte d'un gisement à partir du jour.

4.3.3 Travaux préparatoires en veine Terme générique désignant la réalisation de toutes excavations minières qui subdivisent systématiquement en différents panneaux les gîtes ou parties de ceux-ci, qui ont été reconnus grâce aux travaux de préparation au rocher, afin de les préparer en vue de l'exploitation.

4.3.4 Abattage Détachement du contenu des veines par rapport au massif.

4.3.5 Transport, extraction Terme générique désignant le déplacement des tonnages abattus. Dans les indications statistiques, les limites opératoires doivent être indiquées.

4.3.6 Soutènement Terme générique désignant tous les moyens mis en oeuvre pour maintenir ouvertes et sûres les excavations minières.

4.3.7 Exhaure Tous les processus ou installations et dispositifs utilisés au fond ou au jour afin d'extraire l'eau des excavations minières et d'assurer la collecte, la décantation et la dérivation des venues d'eau (cf. 4.3.15.)

4.3.8 Aérage Ensemble de tous les processus et dispositifs qui ont pour objet d'apporter dans les cavités minières l'air frais nécessaire et de diluer et d'emporter à l'extérieur l'air vicié, toxique, ou les mélanges grisouteux, ainsi que d'améliorer le climat de la mine.

4.3.9 Journées d'extraction Jours ouvrés pendant lesquels l'extraction fonctionne effectivement.

4.3.10 Exploitation à ciel ouvert, exploitation en découverte Exploitation de gisements faisant partie d'une ou plusieurs veines à partir du jour et après déblaiement des terres qui recouvrent les gisements.

4.3.11 Exploitation à ciel ouvert de grande profondeur Exploitation à ciel ouvert dont la profondeur est supérieure à 200 m.

4.1.23 Steinkohleneinheit (SKE) - Einheitsbrennstoff Brennstoff mit einem Heizwert von 29,3 MJ/kg ($\cong$ 7.000 kcal/kg) als mengenmässige (t) Bezugsbasis.

4.1.23 Tonelada equivalente en carbón o t.e.c. Combustible de un p.c.i. de 29,3 MJ/Kg. ($\cong$7.000 Kcal/kg) considerado como referencia cuantitativa.

4.2 Lagerstätten—in Bearbeitung

4.2 Yacimientos—en Proceso de elaboración

4.3 Gewinnung

4.3.1 Bergbau unter Tage Abbau von Lagerstätten nutzbarer Mineralien, die durch Stollen oder Schächte aufgeschlossen und aus sich daran anschliessenden Grubenbauen abgebaut werden.

4.3.2 Ausrichtung Sammelbegriff für die Herstellung aller Grubenbaue, die das Aufsuchen und Anfahren einer Lagerstätte von der Tagesoberfläche aus bezwecken.

4.3.3 Vorrichtung Sammelbegriff für die Herstellung aller Grubenbaue, welche die durch die Ausrichtung aufgeschlossene Lagerstätte oder Teile derselben planmässig in einzelne Baufelder unterteilen, um diese für den Abbau vorzubereiten.

4.3.4 Gewinnung (Lösen) Herausiösen des Lagerstätteninhalts aus dem Gebirgsverband.

4.3.5 Förderung Sammelbegriff für das Fortbewegen der hereingewonnenen Massen. Bei statistischen Angaben sind die Grenzen des Betriebsbereiches anzugeben.

4.3.6 Ausbau Sammelbegriff für alle Mittel, die zum Offenhalten und Sichern von Grubenbauen in diese eingebracht werden.

4.3.7 Wasserhaltung Alle Vorgänge oder Anlagen und Einrichtungen unter oder über Tage zum Fernhalten des Wassers von den Grubenbauen sowie zum Sammeln, Klären und Ableiten der zufliessenden Wässer.

4.3.8 Bewetterung Gesamtheit aller Vorgänge und Einrichtungen, die den Grubenbauen die notwendige Frischluft zuführen, matte, giftige oder schlagende Wetter bis zur Unschädlichkeit verdünnen und abführen und das Grubenklima verbessern.

4.3.9 Fördertage Arbeitstage, an denen tatsächlich gefördert wurde.

4.3.10 Tagebau Abbau von Lagerstätten aus einem oder mehreren Flözen von der Tagesoberfläche aus nach Abräumen des die Lagerstätte überdeckenden Gebirges.

4.3.11 Tieftagebau Tagebau mit einer Tiefe von mehr als 200 m.

4.3 Minería y transporte

4.3.1 Minería subterránea Minería en la que el acceso al yacimiento minero se realiza por medio de pozos y labores subterráneas.

4.3.2 Trabajos de reconocimiento Labores dirigidas a localizar la presencia de yacimientos económicos de carbón y determinar su extensión y calidad.

4.3.3 Trabajos de preparación Labores mineras sistemáticas realizadas para abrir los yacimientos de carbón reconocidos, y preparar su arranque y explotación.

4.3.4 Arranque Separación del contenido de las vetas con relación al macizo.

4.3.5 Transporte, arrastre Término genérico aplicado al movimiento de los productos obtenidos en la mina desde el punto de arranque. En los datos estadísticos deben señalarse los limites de la zona a que se refieren.

4.3.6 Sostenimiento Término genérico referente a todos los medios empleados para mantener abiertas y seguras las excavaciones mineras.

4.3.7 Desagüe Todos los trabajos o instalaciones, tanto de superficie como de interior, cuya finalidad es extraer el agua de las excavaciones mineras y asegurar la recogida, decantación y derivación de las aguas que llegan a la mina.

4.3.8 Ventilación Conjunto de procedimientos e instalciones destinados a suministrar aire puro a las labores mineras, diluir y eliminar el aire viciado y tóxico, asi como las mezclas de grisú y mejorar el clima de la mina.

4.3.9 Días de produccion Días de trabajo en los que, efectivamente, se produce carbón.

4.3.10 Explotación a cielo abierto Es la explotación realizada desde la superficie después de desmontar el recubierto de un yacimiento formado por una o varias capas.

4.3.11 Explotación profunda a cielo abierto Explotaciones a cielo abierto cuya profundidad es superior a 200 metros.

4

4.3.12 Overburden (opencast mining, etc.) All of the layers of rock, parting and coal occurring as mining losses, moved or to be moved from an opencast, etc., mine with a view to winning the coal.

4.3.13 Parting; dirtband; interseam sediments (Aust.) Layers of rock or sediments that occur between useful seams, between strata or in the seam.

4.3.14 Overburden ratio; stripping ratio (opencast, etc., mining)

$$R = \frac{\text{Thickness of overburden}}{\text{Thickness of useful coal}}$$

The linear ratio obtained directly by drilling, or derived from drilling data or from data obtained from other types of working, allowing in each case for admissible mining losses.

4.3.15 Drainage; dewatering (opencast, etc., mining) The collection and removal of ground and surface water from the area of opencast working, together with measures to prevent ingress of water into the area, with a view to ensuring the safety of the workings and improving the quality of the solid fuel produced.

4.3.16 Opencast development; open-pit development; open-cut development; strip mining development All measures taken for preparing for the mining of opencast, etc., coal, apart from exploratory work. Under this heading fall all measures that are required for starting the opencast, etc., operation, for extending it and/or transferring it to another field. Often simply referred to as *overburden removal*.

4.3.17 Excavation and loading The fully mechanised winning of the loose and solid material occurring in opencast, etc., mining, including transfer of the material won to transport.

4.3.18 Dumping; tipping (of overburden in opencast, etc., mining) The depositing of waste material removed.

4.4 Processing and Utilisation

4.4.1 Preparation; treatment; dressing (USA); **processing** (SA); **beneficiation** (SA) Collectively, mechanical, physical and physico-chemical processes applied to raw fuel to make it suitable for a particular use.

4.4.2 Cleaning; separation The treatment of coal to lower the mineral matter (ash) content. In coal cleaning the feed material is separated into the substances of which it is composed on the basis of the heterogeneous physical or physico-chemical characteristics of the differing substances.

4.4.3 Sizing; grading The division of a material into products between nominal size limits using screens, or pneumatic or hydraulic methods.

4.4.4 Size reduction Reducing the granular size of a feed material by crushing, grinding or pulverising; ''crushing'' implies size reduction into relatively coarse particles; ''grinding'' or ''pulverising'' into relatively fine particles.

4.3.12 Morts-terrains, découverte, déblais (exploitation à ciel ouvert) Ensemble des couches de terrains à déplacer ou déplacées pour l'extraction du charbon dans une exploitation à ciel ouvert, des stériles intercalaires et de la fraction du combustible qui constitue les pertes d'exploitation.

4.3.13 Stampe, barre Couches de terrains stériles ou inclusions qui se trouvent entre des veines utilisables, entre plusieurs couches de combustible ou dans une veine.

4.3.14 Rapport (linéaire) de découverture (exploitation à ciel ouvert)

$$R = \frac{\text{épaisseur des morts-terrains}}{\text{épaisseur du charbon utilisable}}$$

Rapport linéaire obtenu directement par sondage ou par tout autre moyen ou dérivé des sondages de reconnaissance, compte tenu des pertes admises à l'abattage.

4.3.15 Drainage (exploitation à ciel ouvert) Ensemble des dispositions prises en vue d'assurer l'écoulement des eaux qui affluent ou qui s'écoulent dans les morts-terrains et dans le front de taille, ainsi que dans le remblai et, en outre, dans la zone environnant l'exploitation, pour obtenir une stabilité suffisante des banquettes et des morts-terrains par retrait des eaux qui sourdent du rocher sur les côtés de la zone d'abattage, ainsi que toutes les mesures en vue de diminuer la teneur en eau de la veine de combustible.

4.3.16 Ouverture d'une mine à ciel ouvert Ensemble des dispositions prises en vue de préparer l'extraction du charbon d'une mine à ciel ouvert à l'exclusion des travaux de reconnaissance. Sont incluses dans cet ensemble toutes les dispositions nécessaires pour permettre le début de l'exploitation de la mine à ciel ouvert, son extension et/ou pour assurer le passage à un autre panneau.

4.3.17 Découverte (exploitation à ciel ouvert) Extraction entièrement mécanisée des terrains meubles qui se trouvent à fleur de terre dans une exploitation à ciel ouvert ainsi que des parties rocheuses, y compris le transport des masses extraites jusqu'aux moyens de transport.

4.3.18 Culbutage (exploitation à ciel ouvert) Mise en tas de déblais.

4.4 Préparation et Valorisation

4.4.1 Préparation Procédés et processus mis en oeuvre pour l'obtention de produits commercialement valorisables à partir de combustibles bruts par des traitements physiques ou physico-chimiques.

4.4.2 Triage, épuration Séparation d'un produit à traiter d'après sa composition en différentes matières, en utilisant les différentes caractéristiques physiques ou physico-chimiques des différentes matières.

4.4.3 Calibrage, classement Séparation d'un produit à traiter d'après les calibres, au moyen d'un ou dans un courant d'air ou d'eau.

4.4.4 Comminution (concassage ou broyage) Consiste à amener un produit à traiter à une granulométrie plus fine, par concassage ou broyage.

4.3.12 Abraum (Tagebau) Teil der Erdrinde, der zur Freilegung und somit zuer Nutzbarmachung eines oder mehrerer Rohstoffkörper im Tagebauraum bewegt werden muss und sich aus dem Deckgebirge, den Mitteln, dem tagebautechnisch bedingten Abtrag von Liegendschichten und den Abbauverlusten zusammensetzt.

4.3.13 Zwischenmittel, Bergemittel Gebirgsschichten oder Einlagerungen, die zwischen nutzbaren Flözen, zwischen mehreren Brennstoffbänken oder im Flöz auftreten.

4.3.14 Betriebliches Mächtigkeitsverhältnis (Tagebau)

$$\frac{A}{K_n} = \frac{\text{Abraummächtigkeit}}{\text{Mächtigkeit der nutzbaren Kohle}}$$

Das aus Bohrungen unmittelbar oder aus diesen oder sonstigen Aufschlüssen abgeleitete lineare Verhältnis unter Berücksichtigung der zulässigen eintretenden linearen Gewinnungsverluste.

4.3.15 Entwässerung (Tagebau) Gesamtheit aller Massnahmen zum Abzug der im Deckgebirge und Lösen, Fassen und Ableiten vom Grund- und Oberflächenwasser im Tagebauraum und Fernhalten vom Grund- und Oberflächenwasser vom Tagebauraum zur Gewährleistung der Tagebausicherheit und der Anbauführung sowie zur Erhöhung der Gebrauchseigenschaften des festen mineralischen Rohstoffes.

4.3.16 Tagebauaufschluss Sämtliche Massnahmen zur Vorbereitung der Gewinnung von Kohle im Tagebau ausser den Erkundungsarbeiten. Hierunter fallen alle Massnahmen, die für den Beginn des Tagebaubetriebes, zu dessen Erweiterung und/oder zum Uebergang in ein anderes Baufeld notwendig sind.

4.3.17 Baggerung Vollmechanisiertes Gewinnen der im Tagebau anstehenden Locker- und Festgesteine einschliesslich der Uebergabe der gewonnenen Massen an die Fördermittel.

4.3.18 Verkippung (Tagebau) Ablagern von Abraum

4.4 Verarbeitung und Verwertung

4.4.1 Aufbereitung Verfahren und Vorgänge zur Herstellung von technisch verwertbaren Erzeugnissen aus Rohbrennstoffen auf physikalischer oder physikalisch-chemischer Grundlage.

4.4.2 Sortierung Trennung eines Aufgabegutes nach seiner stofflichen Zusammensetzung aufgrund der verschiedenartigen physikalischen oder physikalisch-chemischen Eigenschaften der unterschiedlichen Stoffe.

4.4.3 Klassierung Trennung eines Aufgabegutes nach Korngrössen durch Siebe oder im Luft- bzw. Wasserstrom.

4.4.4 Zerkleinerung Ueberführen eines Aufgabegutes in eine feinere Körnung durch Brechen oder Mahlen.

4.3.12 Recubierto, montera (explotación a cielo abierto). Es el conjunto de los estratos de roca que cubren el yacimiento, las intercalaciones de roca y carbón consideradas como estériles que tienen que ser movidas para obtener el combustible útil.

4.3.13 Intercalaciones de estéril Son los estratos de roca estéril que se presentan entre las capas del combustible o dentro de las mismas capas.

4.3.14 Relación de recubierto (explotación a cielo abierto). Es la relación entre el espesor del recubierto y el espesor del carbón útil; esta relación lineal se obtiene directamente de sondeos o por otros medios debiendo considerarse las pérdidas admisibles en la explotación.

4.3.15 Drenaje, desagüe (explotación a cielo abierto) Conjunto de disposiciones tomadas para asegurar el desagüe de las aguas que afluyen o se filtran en las monteras y en el frente de excavación así como en los rellenos y, además, en la zona que rodea la explotación, a fin de obtener una estabilidad suficiente de las bancadas y los recubrimientos, retirando las aguas procedentes de la roca a cotas superiores a la zona de arranque así como todas las medidas a fin de disminuir el contenido en agua de la capa de combustión.

4.3.16 Apertura a cielo abierto, preparacion Conjunto de todos los trabajos preparatorios para la obtención del carbón a cielo abierto, excepto los de reconocimiento. Se incluye en este título los trabajos que se precisan para comenzar la explotación o para su ampliación y/o su traslado a otro campo.

4.3.17 Desmonte, (explotación a cielo abierto) Arranque mecanizado de los materiales compactos o sueltos que se encuentran a flor de tierra en la explotación a cielo abierto, incluido el transporte de los materiales, así obtenidos, hasta el medio de transporte.

4.3.18 Vertido, apilado (del recubierto en explotación a cielo abierto) Depósito del recubierto.

4.4 Preparación y Valoración

4.4.1 Preparación Procedimiento y proceso para la obtención de productos comercialmente valorables, a partir de los combustibles brutos, por medio de tratamientos físicos o fisicoquímicos.

4.4.2 Lavado Separación del material tratado según las distintas substancias que lo componen, en base de las diferencias características físicas y físico-químicas de cada una.

4.4.3 Ciasificación, clasificación granulométrica Separación del material a tratar en varios productos según sus calibres, utilizando una criba o por medio de corrientes de aire o de agua.

4.4.4 Reducción de tamaño (trituracion o molienda) Paso del producto a tratar a otro de granulometrría más fina por medio de una trituración o de una molienda.

4.4.5 Solids/water separation; dewatering The removal of water from wet materials by means other than evaporation. A collective term defining all processes aimed at the mechanical concentration of solids (thickening), at obtaining a low water content product (dewatering) or at obtaining a low solids content product (clarifying). See also 4.4.11.

4.4.6 Solids/gas separation A collective term for all processes aimed at producing a gas low in solids content, with a view to recovering the solids or producing dust-free gas.

4.4.7 Feed regulation; proportioning (USA); **dosing** (SA) Feeding a solid, liquid or gaseous substance by volume or weight.

4.4.8 Blending Mixing in predetermined and controlled quantities to give a uniform product.

4.4.9 Mass yield The amount of useful product obtained from any operation expressed as a percentage of the feed material by weight. The mass yield may be referred to individual preparation operations or to the preparation plant as a whole.

4.4.10 Briquetting; agglomeration The compacting of fuel fines, with or without a binder, in a moulding press to obtain products of the same shape and dimensions.

4.4.11 Drying The lowering of the water content of the solid product by evaporation or vaporization.

4.4.12 Steam drying Thermal drying using steam as the source of heat.

4.4.13 Flue gas drying Thermal drying using flue gases as the source of heat.

4.4.14 Cooling; quenching Removing the heat from a substance in order to lower its temperature, without changing its phase.

4.4.15 Binderless briquetting The shaping and consolidation of the briquetting feedstock using pressure and the cohesive properties of the material.

4.4.16 Briquetting with binders The shaping and consolidation of the briquetting feedstock using pressure and a binding material.

4.4.17 Hot briquetting The briquetting of a fuel of fine granular size using a fuel that softens on heating and acts, therefore, as a binder.

4.4.18 Carbonisation; dry distillation The heating of organic raw materials in the absence of air to obtain coke, crude gas and crude tar.

4.4.19 Low-temperature carbonisation; semi-coking; low-temperature distillation (USA) (SA) The heating of bituminous fuel in the absence of air to a temperature of 500 to 800°C with recovery of low-temperature carbonisation/distillation gas, low-temperature tar and low-temperature coke or semi-coke. In French and German speaking countries the heating temperatures are around 600°C for hard coal, 400-600°C for brown coal, 350-550°C for peat.

4.4.20 High-temperature carbonisation (coking); high-temperature distillation (USA) (SA) The heating of bituminous fuel in the absence of air to a temperature above 800°C with recovery of gas (coke-oven gas, coal gas), high-temperature coke (hard coke, gas coke) and high-temperature tar. In French and German speaking countries the heating temperatures are above 1000°C for hard coal, 900°C for brown coal.

4.4.5 Séparation solides/eau Terme générique pour désigner tous les procédés de concentration mécanique de solides (épaississement), pour obtenir un produit pauvre en eau (égouttage), ou un produit pauvre en solides (clarification) par des procédés mécaniques (Cf. également 4.4.11.).

4.4.6 Séparation solides/gaz Terme générique désignant tous les procédés appliqués en vue d'obtenir un gaz pauvre en solides, dans le but, soit de récupérer les solides, soit d'obtenir un gaz sans poussières.

4.4.7 Dosage Consiste à mettre la quantité convenable d'un corps solide, liquide ou gazeux d'après le volume ou la masse.

4.4.8 Mélange Consiste à réunir physiquement des matières se trouvant dans la même phase ou dans des phases différentes, en faisant en sorte que la répartition des caractéristiques soit identique en tous les points du produit mélangé.

4.4.9 Rendement à la préparation Rapport, exprimé en pourcentage, de la masse (débit en masse) du produit valorisable à la masse (débit en masse) du produit à traiter. Le rendement peut être indiqué pour les différentes opérations de préparation ou pour l'ensemble d'une installation de préparation.

4.4.10 Agglomération Consiste à agglomérer un combustible de fine granulométrie en le moulant par pression, avec ou sans liant, pour obtenir des produits de même configuration et de mêmes dimensions.

4.4.11 Séchage Diminution de la teneur en eau du produit solide par évaporation ou vaporisation.

4.4.12 Séchage à la vapeur Séchage par apport de chaleur au moyen de vapeur.

4.4.13 Séchage au gaz de combustion Séchage par apport de chaleur au moyen de gaz de combustion.

4.4.14 Refroidissement Évacuation de chaleur par un abaissement de température, sans modification de la phase en la matière.

4.4.15 Agglomération sans liant Mise en forme et solidification par emploi de pression et en utilisant les tensions superficielles du produit aggloméré.

4.4.16 Agglomération avec liant Mise en formes et solidification par utilisation de pression et en ajoutant un liant au produit aggloméré.

4.4.17 Agglomération à chaud Agglomération d'un combustible de fine granulométrie avec un combustible ramolli par apport de chaleur et agissant ainsi comme liant.

4.4.18 Carbonisation,. pyrogénation Chauffage de matières organiques brutes en l'absence d'air en vue de la production de coke, de gaz brut et de goudron brut.

4.4.19 Carbonisation à basse température (semi-distillation) Chauffage de combustibles bitumineux en l'absence d'air, à une température comprise entre 450 et 600°C environ, en vue d'extraire les produits de décomposition gazeux et liquides, de sorte que l'on obtient du gaz primaire, du goudron primaire et du semi-coke.

4.4.20 Carbonisation à haute température (cokéfaction) Chauffage de combustibles bitumineux en l'absence d'air à une température supérieure à 900°C, en vue d'extraire les produits de décomposition gazeux et liquides, de sorte que l'on obtient du gaz de four à coke (gaz de cokerie), du coke de haute température et du goudron de haute température.

4.4.5 Feststoff-Wasser-Trennung Sammelbegriff für alle Verfahren zum mechanischen Anreichern von Feststoff (Eindicken), Herstellen eines wasserarmen Erzeugnisses (Entwässern) oder eines feststoffarmen Erzeugnisses (Klären) durch mechanische Vorgänge.

4.4.6 Feststoff-Gas-Trennung Sammelbegriff für alle Verfahren zur Erreichung eines feststoffarmen Gases zum Zwecke der Rückgewinnung des Feststoffes oder der Erzielung staubfreien Gases.

4.4.7 Dosierung Zuteilen eines festen, flüssigen oder gasförmigen Stoffes nach Volumen oder Masse.

4.4.8 Mischung Physikalishces Vereinigen von Stoffen gleicher oder unterschiedlicher Phase mit dem Ziel, dass die betrachteten Eigenschaftsverteilungen an allen Stellen des gemischten Gutes gleich sein sollen.

4.4.9 Masse-Ausbringen Verhältnis der Masse (Massenstrom) an verwertbaren Erzeugnissen zur Masse (Massenstrom) an Aufgabegut in Prozent. Das Masse-Ausbringen kann für einzelne Aufbereitungsvorgänge oder eine gesamte Aufbereitungsanlage angegeben werden.

4.4.10 Brikettierung Stückigmachen eines feinkörnigen Brennstoffes mit Formgebung durch Verpressen mit oder ohne Bindemittel zu Gebilden gleicher Gestalt und Grösse.

4.4.11 Trocknung Verminderung des Wassergehaltes des Feststoffes durch Verdunsten oder Verdampfen.

4.4.12 Dampftrocknung Trocknen durch Wärmezufuhr mittels Dampf.

4.4.13 Feuergastrocknung Trocknen durch Wärmezufuhr mittels Feuergasen.

4.4.14 Kühlung Abführen von Wärme zur Erniedrigung der Temperatur ohne Aenderung der Phase des Stoffes.

4.4.15 Bindemittellose Brikettierung Formgeben und Verfestigen durch Anwendung von Druck unter Nutzung von Oberflächenkräften des Brikettiergutes.

4.4.16 Brikettierung mit Bindemittel Formgeben und Verfestigen durch Anwendung von Druck unter Zugabe von Bindemittel zum Brikettiergut.

4.4.17 Heissbrikettierung Brikettierung eines feinkörnigen Brennstoffes mit einem Brennstoff, der durch Wärmezufuhr erweicht ist und dadurch als Bindemittel wirkt.

4.4.18 Entgasung Erhitzen von organischen Rohstoffen unter Luftabschluss zur Gewinnung von Koks, Rohgas und Rohteer.

4.4.19 Tieftemperatur-Entgasung (Schwelung) Erhitzen bituminöser Brennstoffe unter Ausschluss von Luft auf eine Temperatur von etwa 450 bis 600°C, bei der die hierbei entstehenden gasförmigen und flüssigen Zersetzungsprodukte ausgetrieben werden, so dass Schwelgas, Schwelteer und Schwelkoks entstehen.

4.4.20 Hochtemperatur-Entgasung (Verkokung) Erhitzen bituminöser Brennstoffe unter Auschluss von Luft auf eine Temperatar von über 900°C, bei der bie hierbei entstehenden gasförmigen und flüssigen Zersetzungsprodukte ausgetrieben werden, so dass Koksofengas (Kokereigas), Hochtemperaturteer und Hochtemperaturteer und Hochtemperaturkoks entstehen.

4.4.5 Separación sólido/agua Término genérico para designar todos los procedimientos de concentración mecánica de sólidos (espesamiento) a fin de obtener un sólido con poca agua (agotado) o un agua con poco sólido (aclarado) (ver igualmente 4.4.11).

4.4.6 Separación sólido/gas Término que comprende todos los proceos dirigidos a obtener un gas con poco contenido de sólidos con la finalidad, o bien de recuperar los sólidos arrastrados pór el gas, o bien de obtener un gas sin polvo.

4.4.7 Dosificación Consiste en introducir un material sólido, líquido o gaseoso, en cantidad determinada, bien por el volumen o bien por el peso.

4.4.8 Mezclado Consiste en reunir físicamente materias que se encuentran en la misma fase o en fases diferentes, haciéndolo de forma que la distribución de sus características sea la misma en todas las partes del producto.

4.4.9 Rendimiento en la preparación Relación, expresada en procentaje, entre la masa del producto evaluable y la masa del material a tratar. El rendimiento puede referirse a cada una de las operaciones o bien al conjunto de una instalación de preparación.

4.4.10 Briqueteado, aglomeración Consiste en aglomerar un combustible de granulometría fina mediante un prensado con o sin aglomerante, para obtener productos de igual forma y tamaño.

4.4.11 Secado Disminución del contenido de agua del producto sólido mediante evaporación y vaporización.

4.4.12 Secado por vapor Secado aportando calor por medio de vapor.

4.4.13 Secado por humos Secado aportando calor por medio de los gases de combustión.

4.4.14 Enfriado Evacuación del calor mediante reducción de temperatura sin que se produzca variación de la fase del material.

4.4.15 Briqueteado sin aglomerante Moldeado y consolidación del material menudo tratado, empleando simplemente presión aprovechando las tensiones superficiales del producto aglomerado.

4.4.16 Briqueteado con aglomerante Moldeado y consolidación del material menudo, empleando presión y un producto aglomerante.

4.4.17 Briqueteado en caliente Aglomerado de un combustible de granulometría fina empleando otro combustible que se reblandece por el calor y actua como aglomerante.

4.4.18 Carbonización, pirogenación Calentamiento de materias orgánicas brutas, fuera del contacto con el aire, para obtener coque, gas bruto y alquitrán.

4.4.19 Carbonización a baja temperatura (semi-destilacion) Calentamiento de combustibles bituminosos, fuera del contacto con el aire, a una temperatura entre 450 y 600° C a fin de extraer los productos de descomposición gaseosa y líquidos de manera que se obtiene gas primario, alquitrán primario y semi-coque.

4.4.20 Carbonización a alta temperatura, coquizacion Calentamiento del carbón bituminoso, fuera del contacto con el aire, a temperatura superior a 900° C a fin de extraer los productos de descomposición gaseosos y líquidos, de manera que se obtiene gas de horno de coque (gas de coquería), coque de alta temperatura y alquitrán de alta temperatura.

4

4.4.21 Condensation The conversion of steam, vapour or gas to liquid.

4.4.22 Gas cleaning; gas purification "Gas cleaning" generally implies the removal of particulate matter from gases; "gas purification" generally implies the removal of gaseous impurities from gases.

4.4.23 Coke quenching The removal of the sensible heat from coke discharged from retorts, etc., by means of water (wet quenching) or inert gases (dry quenching).

4.4.24 Heating gas; underfiring gas Gas used to heat coke-ovens, retort settings or other installations in which endothermic reactions take place. According to the type of plant a high CV gas (e.g. coke-oven gas), a low CV gas (e.g. blast-furnace gas) or liquefied petroleum gases are utilised.

4.4.25 Coke yield The amount of coke obtained in the carbonisation process expressed as a percentage of the feed material by weight. Reference conditions should be stated.

4.4.26 Gas yield The volume of gas obtained in the carbonisation process or gasification process expressed as m^3/te (or ft^3/ton) of the fuel feed. Reference conditions should be stated.

4.4.27 Gasification A process of manufacturing fuel gases by reacting the solid fuel with a gasification medium, i.e. air, oxygen or steam.

4.4.28 Liquefaction The conversion of solid fuels into liquid hydrocarbons and related compounds by hydrogenation, by synthesis of gases derived from solid fuels or by solvent extraction of solid fuels.

4.4.29 Underground gasification; in situ gasification (USA) The gasification of the coal in the seam.

4.4.30 Combustion The exothermic reaction of a fuel with oxygen that remains self-supporting once ignition temperature has been achieved.

4.4.31 Grate firing; fixed bed combustion The combustion of medium and large sized fuel supported on a stationary or movable grate.

4.4.32 Pulverised fuel firing with dry ash removal; dry pulverised fuel firing The combustion of pulverised fuel in suspension in the furnace, the temperature of combustion being below that at which the ash fuses.

4.4.33 Pulverised fuel firing with slagging ash removal; slag tap firing The combustion of pulverised fuel in suspension in the furnace, the temperature of combustion being above that at which the ash fuses.

4.4.34 Flue-gas The gas produced by combustion (e.g. CO_2, H_2O, SO_2) and residual gases from the combustion air (N_2, O_2) together with entrained solids and tar fog.

4.4.35 Fly ash; flue dust Solid, particulate material contained in flue-gases, exhaust gases, etc.

4.4.36 Ash (combustion residue) The residue remaining after combustion whose origin is the mineral impurities contained in the fuel; ash may also contain unburned fuel.

Note In fuel analysis ash is defined as the inert residue when the fuel has been completely burned.

4.4.37 Slag; clinker Combustion residue that has melted during combustion of the fuel at a temperature above ash fusion temperature and has resolidified on cooling.

4.4.21 Condensation Transformation de vapeurs ou de gaz en liquides.

4.4.22 Epuration du gaz Opération ayant pour but d'éliminer des fractions du gaz susceptibles d'être valorisées ou gênantes, par exemple : des gaz combustibles, des fumées, des vapeurs (voir également section 6).

4.4.23 Extinction du coke Évacuation de la chaleur sensible du coke qui vient d'être défourné au moyen d'eau (extinction humide) ou au moyen de gaz non combustibles (extinction à sec).

4.4.24 Gaz de chauffage Gaz assurant le chauffage des fours à coke ou de fours à semi-coke ou d'autres ensembles dans lesquels s'effectuent des réactions thermiques. On utilise pour cela, selon le système de four, un gaz riche (en général du gaz de cokerie) ou un gaz pauvre (par exemple: du gaz de haut-fourneau) ou des gaz de pétrole liquéfiés.

4.4.25 Rendement en coke Masse de coke produite par la carbonisation rapportée à la masse du produit enfourné et exprimée en pour cent. Les conditions de référence doivent être indiquées dans chaque cas.

4.4.26 Rendement en gaz Volume de gaz obtenu par la carbonisation ou par la gazéification, rapporté à la masse de combustibles mis en oeuvre et exprimé en m^3/t. Les conditions de references doivent être indiquées dans chaque cas.

4.4.27 Gazéification Procédé de fabrication de gaz ou combustibles par réaction de combustibles solides avec un agent de gazéification, par exemple : de l'air ou de l'oxygène et de la vapeur d'eau (voir également section 6).

4.4.28 Liquéfaction Conversion du combustible solide en hydrocarbure liquide et composés asscoiés par hydrogénation, par synthèse de gaz dérivés du combustible solide ou par extraction de solvants.

4.4.29 Gazéification souterraine Gazéification de charbon dans le gisement.

4.4.30 Combustion Réaction d'un combustible avec de l'oxygène, avec dégagement de chaleur qui apparaît spontanément lorsque la température d'inflammabilité est atteinte.

4.4.31 Combustion en couche, chauffe sur grille Combustion d'un combustible en gros morceaux sur une grille fixe ou mobile.

4.4.32 Chauffe au charbon pulvérisé avec extraction à sec des cendres Combustion d'un combustible pulvérise dans une chambre de combustion. La température de combustion est inférieure à la température de ramollissement des cendres.

4.4.33 Chauffe au charbon pulvérisé à fusion de cendres Combustion d'un combustible pulvérisé dans une chambre de combustion. La température de combustion est supérieure à la température de fusion des cendres.

4.4.34 Fumées Les gaz produits par la combustion (par exemple, CO_2, H_2O, SO_2) et les gaz provenant de l'air de combustion (N_2, O_2) ainsi que les matières solides et le brouillard de goudron entraînés.

4.4.35 Poussières, cendres volantes Matières solides contenues dans les fumées, les gaz résiduaires ou les vapeurs.

4.4.36 Cendres (résidus de combustion) Résidus de combustion qui proviennent des impuretés minérales contenues dans le combustible, elles peuvent également contenir du combustible non brûlé.

4.4.37 Mâchefers (sous produits) Résidus de combustion qui ont d'abord été liquéfiés lorsque la température de fusion des cendres a été dépassée et qui ont été ensuite resolidifiés par refroidissement.

4.4.21 Kondensation Umwandeln von Dämpfen und Gasen in Flüssigkeiten.

4.4.22 Gasreinigung Verfahren zum Entfernen gewinnungswürdiger oder störender Bestandteile aus Gasen, z.B. Brenngasen, Rauchgasen, Brüden.

4.4.23 Kokskühlung Abführung der fühlbaren Wärme des frischerzeugten Kokses durch Wasser (Löschung) oder durch nicht brennbare Gase (Trockenkühlung).

4.4.24 Unterfeuerungsgas Gas zur Beheizung von Koks- oder Schwelöfen oder anderen thermischen Reaktionsaggregaten. Je nach Ofensystem kommen hierfür Starkgas (meist Koksofengas) oder Schachgas (z.B. Gichtgas) oder Flüsiggas in Schwachgas

4.4.25 Koksausbringen Masse an Koks, die bei der technischen Entgasung anfällt, bezogen auf die Masse des Einsatzgutes und in Prozent ausgedrückt. Die Bezugszustände sind jeweils anzugeben.

4.4.26 Gasausbringen Volumen an Gas, das bei der technischen Entgasung oder Vergasung anfällt, bezogen auf die Masse des eingesetzten Brennstoffes, ausgedrückt in m^3/t. Die Bezugszustände sind jeweils anzugeben.

4.4.27 Vergasung Verfahren zur Herstellung von Brenngasen durch Reaktion fester Brennstoffe mit einem Vergasungsmittel, z.B. Sauerstoff (Luft) oder Wasserdampf.

4.4.28 Verflüssigung Umwandlung von festen Brennstoffen in flüssige Brennstoffe und verwandte Zusammensetzungen durch Hydrierung, durch Synthese von Gasen aus festen Brennstoffen oder durch Entziehen von Lösungsmitteln.

4.4.29 Untertagevergasung Vergasung von Kohlen in ihrer Lagerstätte.

4.4.30 Verbrennung Reaktion von Brennstoffen mit Sauerstoff unter Wärmeentwicklung, die nach Erreichen der Entzündungstemperatur selbsttätig in Gang bleibt.

4.4.31 Verbrennung in der Schicht (Rostfeuerung) Verbrennung von grobkörnigen Brennstoffen auf einem festen oder bewegten Rost.

4.4.32 Verbrennung im Flug (Staubfeuerung) mit trockenem Ascheabzug Verbrennung von staubförmigem Brennstoff im Verbrennungsraum. Die Verbrennungstemperatur liegt unterhalb der Ascheerweichungstemperatur.

4.4.33 Verbrennung im Flug (Staubfeuerung) mit flüssigem Ascheabzug Verbrennung von staubförmigem Brennstoff im Verbrennungsraum. Die Verbrennungstemperatur liegt oberhalb der Ascheschmelztemperatur.

4.4.34 Rauchgas Bei der Verbrennung entstehende (z.B. CO_2, H_2O, SO_2) und aus der Verbrennungsluft verbliebene Gase (N_2, O_2) sowie mitgerissene Feststoffe und Teernebel.

4.4.35 Flugstaub (Flugasche) In Rauchgasen, Abgasen oder Brüden enthaltener Feststoff.

4.4.36 Asche (Feuerungsrückstand) Verbrennungsrückstand, der aus den im Brennstoff enthaltenen mineralischen Begleitstoffen entstanden ist. Er kann auch unverbrannten Brennstoff enthalten.

4.4.37 Schlacke (Klinker) Bei der Verbrennung oberhalb der Ascheschmelztemperatur verflüssigter und nach Abkühlung wiederverfestiger Verbrennungsrückstand.

4.4.21 Condensación Transformación de vapores o gases en líquidos.

4.4.22 Depuración de gases Tratamiento de los gases para separar productos aprovechables o perjudiciales, por ejemplo: gas combustible, humos, vapores (veáse, asimismo sección 6).

4.4.23 Apagado de coque Evacuación del calor sensible del coque a su salida del horno por agua (extinción húmeda) o por gases incombustibles (extinción en seco).

4.4.24 Gas de caldeo Gas empleado en el calentamiento de los hornos de coque o semicoque o de otras cámaras en las que se producen reacciones térmicas.
Según el tipo de hornos puede tratarse de gas rico (en general gas de coque), gas pobre (por ejemplo gas de horno alto) o gas licuado de petróleo.

4.4.25 Rendimiento en coque Cantidad de coque obtenida en el proceso de coquización expresada en porcentaje en peso del material introducido en el horno. En cada caso deben de indicarse las condiciones en que se realiza.

4.4.26 Rendimiento en gas Volumen de gas obtenido en la carbonización o en la gasificación expresado en m^3/T del combustible introducido. Debe indicarse las condiciones en que se mide el gas.

4.4.27 Gasificación Es el proceso de fabricación de gases combustibles mediante la reacción de un combustible sólido con un medio gasificante como aire, oxigeno o vapor de agua, (vease igualmente sección 6).

4.4.28 Licuefacción Procedimiento de conversión de combustibles sólidos en hidrocarburos líquidos y compuestos derivados, por síntesis de gases derivados de combustibles sólidos o por extracción por disolventes.

4.4.29 Gasificación subterránea Es la gasificación del carbón en su propio yacimiento.

4.4.30 Combustión Reacción de un combustible con el oxígeno, con desprendimiento de calor que aparece espontáneamente al alcanzarse la temperatura de inflamabilidad.

4.4.31 Combustión en parrilla Combustión del carbón en granos de tamaño medio o grande soportado por una parrilla fija o móvil.

4.4.32 Combustible de carbón pulverizado con eliminación de las cenizas en seco Combustión de un combustible pulverizado en suspensión en la cámara de combustión. La temperatura de combustión es inferior a la de fusión de las cenizas.

4.4.33 Combustión del carbón pulverizado con eliminación de cenizas fundidas Combustión del carbón pulverizado en suspensión en la cámara de combustión. La temperatura de combustión es superior a la defusión de las cenizas.

4.4.34 Humos (plural) Los gases resultantes de la combustión (por ejemplo CO_2, H_2O, SO_2) y los procedentes del aire de combustion $(N_2O_2)^2$ asi como las partículas sólidas y alquitranes que arrastra.

4.4.35 Cenizas volantes Son las partículas sólidas contenidas en los humos, gases residuales y vapores.

4.4.36 Cenizas (residuos de la combustión) Residuo de la combustión que proviene de las impurezas minerales que contiene el combustible. Puede contener combustible inquemado.

4.4.37 Escorias (sub-productos) residuos de la combustión que por tener una temperatura de fusión inferior a la de combustion, fueron fundidas y, por posterior enfriamiento, se han solidificado inmediatamente.

4

4.5 Properties

4.5.1 Size distribution; size consist A collective term for data on the quantitative distribution of granular sizes occurring in aggregate material. The proportion of various sizes in a product.

4.5.2 Bulk density The mass per unit volume occupied by a collection of fuel particles; the conditions under which the measurement is made should be stated, e.g. in the case of compacted bulk density.

4.5.3 Moisture content The water content of the moist fuel, expressed as percentage by weight.

Total moisture The sum of the free moisture and the inherent moisture.

Moisture in the analysis sample The moisture in the fuel sample at the time of the laboratory analysis.

Moisture-holding capacity The moisture in the fuel sample after it has attained equilibrium with the air to which it is exposed, standard conditions being 30°C and 96 to 97% relative humidity.

4.5.4 Ash content The residue of combustion obtained when the fuel is incinerated at a temperature of 815°C under specified conditions, expressed as percentage by weight.

4.5.5 Inerts content The sum of the constituents of a fuel which decrease its efficiency in use, e.g. mineral matter (ash) and moisture in fuel for combustion, or fusain in coal for carbonisation, expressed as a percentage of the fuel as received, by weight.

4.5.6 Sulphur content; sulfur content The sulphur in the fuel expressed as a percentage by weight.

4.5.7 Tar yield The amount of tar obtained in a standardised dry distillation test method, expressed as a percentage of the feed material by weight.

4.5.8 Gross calorific value; gross heating value (USA, Can.); gross specific energy The number of heat units measured as being liberated when unit mass of fuel is burned in oxygen saturated with water vapour in a bomb under standardised conditions, the residual materials being taken as gaseous oxygen, carbon dioxide, sulphur dioxide and nitrogen, *liquid water* in equilibrium with its vapour and saturated with carbon dioxide, and ash. The standard conditions are defined in ISO-R-1928-71. The international reference temperature is 25°C.

Note In some countries different reference temperatures are applied.

4.5.9 Net calorific value; net heating value (USA, Can.); net specific energy The number of heat units measured as being liberated when unit mass of fuel is burned in oxygen saturated with water vapour in a bomb under standardised conditions, the residual materials being taken as gaseous oxygen, carbon dioxide, sulphur dioxide and nitrogen, *water as water vapour* and ash. The standardised conditions are defined in ISO-R-1928-71. The international reference temperature is 25°C.

Note In some countries different reference temperatures are applied.

4.5.10 Volatile matter (VM) The loss in mass, corrected for moisture, when fuel is heated out of contact with air under standardised conditions, which are defined in ISO R562, the loss being due to the release in the gaseous state of the products of decomposition of the organic matter in the fuel; the residue is termed a *coke button*. The temperature to which the fuel is to be heated is specified as 900°C.

Note In some countries different reference temperatures are applied.

4.5 Caractéristiques

4.5.1 Répartition par grosseur (granulométrie) Expression résumée indiquant la proportion des différents calibres présents dans un produit.

4.5.2 Densité en vrac Masse de l'unité de volume du combustible mis en tas selon des conditions déterminées, qui doivent être précisées.

4.5.3 Humidité, teneur en eau Proportion de l'eau contenue dans un combustible, indiquée en pour cent massique et rapportée au combustible hydraté.
— **Humidité totale** c'est la somme de l'eau superficielle et de l'eau de constitution.
— **Humidité à l'analyse** C'est la teneur en eau du combustible déterminée au moment de l'analyse.
— **Pouvoir de rétention** C'est la teneur en eau du combustible déterminée à l'équilibre à une température de 30°C et pour une humidité relative de l'air de 96 à 97%.

4.5.4 Taux de cendres, teneur en cendres Résidus de combustion récoltés lorsque le combustible est calciné à une température de 815°C, dans des conditions déterminées.

4.5.5 Teneur en inertes Somme de la teneur en humidité totale et de la teneur en cendres considérées dans le combustible hydraté et avec ses cendres.

N.B. : par opposition, on définit dans certains pays ''le charbon pur''.

4.5.6 Teneur en soufre Proportion en masse du soufre contenu dans le combustible.

4.5.7 Rendement en goudron Quantité de goudron obtenue en appliquant un mode opératoire déterminé de distillation sèche et rapportée à l'unité de masse du produit mis en oeuvre.

4.5.8 Pouvoir calorifique supérieur (ou p.c.s.) Chaleur fournie par la combustion complète d'une unité de masse de combustible pour produire du bioxyde de carbone à l'état gazeux, de l'anhydride sulfureux à l'état gazeux et de l'azote à l'état gazeux, ainsi que de l'eau à *l'état liquide* et des cendres, lorsque la température avant la combustion, ainsi que la température des produits obtenus après la combustion est de 25°C (d'après les spécifications internationales ISO-R-1928-71).

N.B.: Dans certains pays, cette température de référence est différente.

4.5.9 Pouvoir calorifique inférieur (ou p.c.i.) Chaleur fournie par la combustion complète d'une unité de masse de combustible pour produire du bioxyde de carbone à l'état gazeux, de l'anhydride sulfureux à l'état gazeux et de l'azote à l'état gazeux, ainsi que de l'eau à *l'état de vapeur* et des cendres, lorsque la température avant la combustion, ainsi que la température des produits obtenus après la combustion est de 25°C (d'après les spécifications internationales ISO-R-1928-71).

N.B.: Dans certains pays, cette température de référence est différente.

4.5.10 Matières volatiles Les matières volatiles (M.V.) sont les produits de la décomposition de la substance organique du combustible qui se dégagent à l'état gazeux par chauffage des combustible solides dans des conditions déterminées, à une température de 900°C. Le résidu recueilli est appelé ''bouton de coke'' (ISO R 562).

N.B. : Dans certains pays, cette température de référence est différente.

4.5 Eigenschaften

4.5.1 Korngrössenverteilung (Körnangsaufbau) Zusammenfassender Begriff für die Angabe der Mengenanteile der in einem Haufwerk vorkomenden Korngrössen.

4.5.2 Schüttdichte Masse der Volumeneinheit des nach festgelegten Bedingungen geschütteten Brennstoffs. Die Bedingungen sind anzugeben.

4.5.3 Wassergehalt Anteil des im Brennstoff enthaltenen Wassers, angegeben in Masseprozenten, bezogen auf den wasserhaltigen Brennstoff.

Gesamtwasser ist die Summe von Oberflächenfeuchtigkeit und innerer Feuchtigkeit.

Analysenfeuchtigkeit ist das Wasser des analysenfeinen Brennstoffes zum Zeitpunkt der Analyse.

Wasserhaltevermögen ist der Wassergehalt des Brennstoffes nach Eintritt der Massekonstanz bei einer Temperatur von 30°C und 96 bis 97% relativer Luftfeuchte.

4.5.4 Aschegehalt Verbrennungsrückstand, der beim Glühen des Brennstoffes unter festgelegten Bedingungen bei einer Temperatur von 815°C entsteht.

4.5.5 Ballastgehalt Summe aus Gesamtwassergehalt und Aschegehalt, bezogen auf den wasser- und aschehaltigen Brennstoff.

Nota: Im gegensatz dazu wird in einigen Ländern die *Reine Kohle* definiert.

4.5.6 Schwefelgehalt Masseanteil des im Brennstoff enthaltenen Schwefels

4.5.7 Teerausbeute Die nach einem bestimmten Untersuchungsverfahren der Trockendestillation entstehende Menge an Teer, bezogen auf die Masseneinheit des Einsatzgutes.

4.5.8 Brennwert (Verbrennungswärme, Verbrennungswert) Wärme, die bei vollständiger Verbrennung einer Masseneinheit des Brennstoffs zu gasförmigen Kohlendioxid, Schwefeldioxid und Stickstoff sowie *flüssigem* Wasser und Asche abgegeben wird, wenn die Temperaur vor der Verbrennung und auch die der entstandenen Produkte nach der Verbrennung 25°C beträgt. (ISO-R 1928-71). In anderen Ländern ist diese Referenztemperatur anders festgelegt.

4.5.9 Heizwert Wärme, die bei vollständiger Verbrennung einer Masseneinheit des Brennstoffs zu gasförmigem Kohlendioxid, Schwefeldioxid und Stickstoff sowie Wasser*dampf* und Asche abgegeben wird, wenn die Temperatur vor der Verbrennung und auch die der entstandenen Produkte nach der Verbrennung 25°C beträgt. (ISO-R 1928-71). In anderen Ländern ist diese Referenztemperatur anders festgelegt.

4.5.10 Flüchtige Bestandteile Flüchtige Bestandteile sind die beim Erhitzen fester Brennstoffe unter festgelegten Bedingungen bei einer Temperatur von 900°C (ISO-R 562) gasförmig entweichenden Zersetzungsprodukte der organischen Brennstoffsubstanz. Der verbleibende Rückstand wird als Tiegelkoks bezeichnet. In manchen Ländern ist diese Referenztemperatur anders festgelegt.

4.5 Características

4.5.1 Distribución por tamanos (granulometría) Término genérico que indica la proporción de los diferentes calibres presentes en un producto.

4.5.2 Densidad aparente Masa de la unidad de volumen del combustible envasado en determinadas condiciones que deben precisarse.

4.5.3 Humedad, contenido de agua Proporción de agua en un combustible húmedo, dado en porcentaje de masa y referido al combustible con humedad.

"Humedad total" Suma del agua superficial y del agua de constitución.

"Humedad de análisis" Es la que contiene la muestra del combustible en el momento del análisis.

"Poder de retención" Es la humedad del combustible determinada en estado de equilibrio a una temperatura de 30° C y para una humedad relativa del aire de 96 a 97%.

4.5.4 Contenido de cenizas Residuo sólido incombustible que queda después de la calcinación del combustible a una temperatura de 815° C en condiciones determinadas.

4.5.5 Contenido de inertes Suma de la humedad total y del contenido de cenizas consideradas en el combustible hidratado y con sus cenizas.

Nota. Por oposición en ciertos casos se define el "carbón puro".

4.5.6 Contenido de azufre Proporción de azufre del combustible expresado en porcentaje de masa.

4.5.7 Rendimiento en alquitrán Cantidad de alquitrán obtenida por un método normalizado de destilación seca expresado como porcentaje en masa del material ensayado.

4.5.8 Poder calorífico superior o.p.c.s. Calor suministrado por la combustión completa de una unidad de masa de combustible para producir anhidrido carbónico, anhidrido sulfurosos en estado gaseoso y nitrógeno en estado gaseoso, así como *agua en estado líquido* y cenizas, siendo la temperatura antes de la combustión, así como la temperatura de los productos obtenidos después de la combustión 25°C de acuerdo con las especificaciones internacionales.

Nota. En algunos países esta temperatura de referencia es diferente.

4.5.9 Poder calorífico inferior Calor suministrado por la combustión completa de una unidad de masa de combustible para producir anhídrido carbónico, anhídrido sulfuroso en estado gaseoso y nitrógen en estado gaseoso, asi como *vapor de agua* y cenizas, siendo la temperatura anes de la combustión, así como la temperatura de los productos obtenidos después de la combustión 25° C, de acuerdo con las especificaciones internacionales.

4.5.10 Materias volátiles Las materias volátiles son los productos de la descomposición de la sustancia orgánica del combustible que se desprende en estado gaseoso por calentamiento de los combustibles sólidos en determinadas condiciones a una temperatura de 900° C (ISO R 562). El residuo obtenido se denomina "botón de coque".

Nota. En algunos países esta temperatura de referencia es diferente.

4

4.5.11 Compressive strength In the case of briquettes formed in a plunger press, the maximum test pressure at the working face of a test plunger before a briquette sample is crushed, under standardised conditions; and, in the case of briquettes formed in a ringroll press, a measure of the strength of the briquette when subjected to a force applied through an approximately pointed element until fracture, under standardised conditions.

4.5.12 Abrasion index; hardness index by standard drum test; abrasion resistance; abrasion strength A measure of the size degradation and abrasion of solid fuels that occur when the fuel is rotated in a test drum under standardised conditions involving mechanical stress. Various indices may be employed, e.g. Micum index, Cockrane abrasion index.

4.5.13 Shatter index; shatter strength A measure of the resistance to size degradation when the fuel is subjected to impact on falling, under standardised conditions. The Shatter index is the percentage of the fuel retained on a sieve of stated aperture after being subjected to the shatter test.

4.5.14 Ash fusibility The behaviour of pelletised ash test samples when heated under standardised conditions. The following characteristic temperatures are identifiable: deformation temperature, hemisphere temperature and flow temperature.

4.5.15 Ash composition The percentage of specific elements in the ash expressed as oxides and acid residues. Examination of ash is generally limited to the determination of silicon, aluminium, iron, calcium, magnesium, potassium, sodium, titanium, phosphorus and sulphur.

4.5.16 Mineral matter content The sum of the inorganic constituents of moisture-free fuel expressed as a percentage by weight.

4.5.17 Proximate analysis The analysis of fuel expressed in terms of moisture, volatile matter, ash and fixed carbon.

4.5.18 Ultimate analysis; elementary analysis The analysis of fuel expressed in terms of its carbon, hydrogen, nitrogen, sulphur and oxygen contents.

4.5.19 Grindability A measure of the energy expended under standardised conditions in order to grind the fuel to a required degree of fineness.

4.5.20 Swelling number; swelling index A measure of the caking properties of coals. The number is obtained by comparing the profile of the sample under test with the profiles of a series of standard cokes under specified conditions.

4.5.21 Swelling behaviour; dilatation The change in volume (contraction, dilation) of a pelletised coal sample when progressively heated under standard conditions. *Dilatation* is the term employed in the UN International Classification of Hard Coals by Type.

4.5.22 Salt content; sodium content (Can.); **alkali content** (Can.) The water soluble salts content of the solid fuel. The analytical method must be stated.

4.5.11 Résistance à la compression-résistance ponctuelle
— **Résistance à la compression (pour les agglomérés fabriqués à la presse à plongeur)** La pression d'essai maximale rapportée à la surface utile d'un plongeur d'essai pour laquelle la cassure d'un échantillon d'aggloméré se produit dans des conditions déterminées.
— **Résistance ponctuelle (pour les agglomérés fabriqués à la presse à cylindre)** Résistance de l'aggloméré auquel on applique une force de façon approximativement ponctuelle jusqu'à cassure de l'aggloméré et dans des conditions déterminées.

4.5.12 Indice de résistance au tambour, résistance à l'abrasion Grandeur de mesure de la désagrégation et de l'abrasion de combustibles solides qui se produisent dans des conditions déterminées sous l'action des contraintes mécaniques subies dans un tambour d'essais.

4.5.13 Indice de résistance au choc, à la chute Grandeur de mesure de la résistance au bris dans le cas où le combustible est soumis à un essai de chute dans des conditions déterminées.

4.5.14 Fusibilité des cendres Comportement d'échantillons comprimés de cendres au cours d'un chauffage appliqué dans des conditions déterminées. On repère les températures caractéristiques suivantes : température de ramollissement, température de goutte et température de fluidification.

4.5.15 Composition des cendres Proportion d'éléments déterminés dans les cendres, exprimés en oxydes et en restes acides. L'examen des cendres se limite en général à la détermination du silicium, de l'aluminium, du fer, du calcium, du magnésium, du potassium, du sodium, du titane, du phosphore et du soufre.

4.5.16 Teneur en matières minérales Proportion totale des composants inorganiques dans le combustible sec.

4.5.17 Analyse immédiate Détermination des teneurs en eau, cendres et matières volatiles.

4.5.18 Analyse élémentaire Teneur de la substance organique du combustible en carbone, hydrogène, oxygène, soufre et azote.

4.5.19 Aptitude au broyage Valeur mesurée de l'énergie qu'il est nécessaire de dépenser, dans des conditions déterminées, pour obtenir un broyage fin d'un combustible.

4.5.20 Indice de gonflement Valeur mesurée du pouvoir agglutinant des houilles. C'est le numéro du profil type (type standard) dont la forme est la plus proche de celle du résidu cokéfié préparé dans des conditions déterminées.

4.5.21 Courbe dilatométrique, dilatabilité Modification du volume (contraction, dilatation) d'un échantillon comprimé de houille en chauffe progressive, dans des conditions déterminées.

4.5.22 Teneur en sels Teneur du combustible solide en sels solubles dans l'eau. Le mode opératoire doit être indiqué.

4.5.11 Druckfestigkeit - Punktfestigkeit (bei Briketts aus Stempelpressen) Die auf die Druckfläche eines Prüfstempels bezogene maximale Prüfkraft, die bei unter festgelegten Bedingungen durchgeführtem Zerdrücken einer Brikettprobe auftritt.

Punktfestigkeit (bei Briketts aus Walzenpressen) Mass für die Festigkeit der Briketts, auf die unter festgelegten Bedingungen bei annähernd punktförmiger Auflage eine Kraft bis zum Bruch einwirkt.

4.5.12 Trommelfestigkeit bzw. Abriebfestigkeit Mass für den Kornzerfall und Abrieb von festen Brennstoffen, die unter festgelegten Bedingungen bei der mechanischen Beanspruchung in einer Prüftrommel entstehen.

4.5.13 Sturzfestigkeit Mass für den Widerstand gegen Kornzerfall, wenn der Brennstoff unter festgelegten Bedingungen durch Fall geprüft wird.

4.5.14 Ascheschmelzverhalten Verhalten von gepressten Ascheprobekörpern beim Erhitzen unter festgelegten Bedingungen. Es werden folgende charakterischen Temperaturen bestimmt : Erweichungs-, Halbkugel- und Fliess-Temperatur.

4.5.15 Aschezusammensetzung Anteile bestimmter Elemente in der Asche, ausgedrückt als Oxide und Säurereste. Die Untersuchung der Asche erstreckt sich im allgemeinen auf die Bestimmungen von Silizium, Aluminium, Eisen, Calcium, Magnesium, Kalium, Natrium, Titan, Phosphor und Schwefel.

4.5.16 Mineralstoffgehalt Summe der anorganischen Bestandteile im wasserfreien Brennstoff.

4.5.17 Kurzanalyse Bestimmung der Gehalte an Wasser, Asche und flüchtigen Bestandteilen.

4.5.18 Elementaranalyse Bestimmung der Gehalte der organischen Brennstoffsubstanz an Wasser, Asche und flüchtigen Bestandteilen (Kohlenstoff, Wasserstoff, Sauerstoff, Schwefel und Stickstoff).

4.5.19 Mahlbarkeit Mass für den zur Feinmahlung eines Brennstoffs unter festgelegten Bedingungen erforderlichen Arbeitsaufwand.

4.5.20 Blähzahl (Blähgrad) Mass für das Backvermögen von Steinkohlen. Sie ist die Nummer desjenigen Musterprofils (Standardtypes), das der Form des Verkokungsrückstandes, hergestellt unter festgelegten Bedingungen, am nächsten kommt.

4.5.21 Dilatationsverlauf Volumenänderung (Kontraktion, Dilatation) eines Steinkohlenpresslings bei fortschreitender Erhitzung unter festgelegten Bedingungen.

4.5.22 Salzgehalt Gehalt des festen Brennstoffes an wasserlöslichen Salzen. Das Analysenverfahren ist anzugeben.

4.5.11 Resistencia a la compresión, resistencia puntual *Resistencia a la compresión (para los aglomerados fabricados en prensa de pistón).*

Presión máxima de ensayo, referida a la superficie útil de un pistón de ensayo, que provoca la rotura, en determinadas condiciones de una probeta de aglomerado.

Resistencia puntual (para los aglomerados fabricados en prensa de cilindros).

Medida de la resistencia del aglomerado al que se aplica una fuerza en forma aproximadamente punctual, hasta su rotura, en condiciones determinadas.

4.5.12 Indice de resistencia al tambor, resistencia a la abrasión Medida de la disgregación y de la abrasión de combustibles sólidos provocadas en determinadas condiciones bajo la acción de tensiones mecánicas en un tambor de ensayo.

4.5.13 Indice de resistencia al choque, a la caída Medida de la resistencia a la fractura en el caso en que el combustible esté sometido a una ensayo de caída en determinadas condiciones.

4.5.14 Fusibilidad de las cenizas Comportamiento de muestras comprimidas de las cenizas sometidas a un calentamiento en condiciones determinadas. Se distinguen las temperaturas características siguientes: temperatura de reblandecimiento, temperatura de fusión y temperatura de fluidez.

4.5.15 Composición de las cenizas Proporción en que se encuentran determinados elementos en las cenizas, expresado en óxidos y en restos ácidos. El examen de las cenizas se limita en general a la determinación de silicio, aluminio, hierro, calcio, magnesio, potasio, sodio, titanio, fósforo y azufre.

4.5.16 Contenido en materia mineral Proporción total de los componentes inorgánicos en el combustible seco.

4.5.17 Análisis inmediato Determinación de los contenidos de humedad, materias volátiles y cenizas.

4.5.18 Análisis elemental Contenido de sustancia del combustible en carbón, hidrógeno, oxígeno, azufre y nitrógeno.

4.5.19 Triturabilidad Medida de la energía necesaria, en condiciones determinadas, para obtener la trituración fina de un combustible.

4.5.20 Indice de hinchamiento Es una medida del poder aglomerante de las hullas. El número se obtiene comparando el perfil de la muestra, después de ser calentada progresivamente en condiciones normalizadas, con una serie de perfiles-tipos numerados.

4.5.21 Curva dilatométrica Variaciones de volumen (contracciones y dilataciones) de una muestra comprimida de hulla calentada progresivamente en condiciones determinadas.

4.5.22 Contenido en sales Contenido de sales, solubles en agua, de un combustible sólido. Deberá indicarse el método de análisis empleado.

4

4.6 Storage

4.6.1 Stock; store; storage yard; storage ground A permanent stock at the surface of the mine of the marketable products of mining, maintained in order to regulate the quantities and even out the characteristics of the products, and also to serve as a reserve and to maintain a balance between production and consumption, together with the transport facilities for supply to and delivery from the stock.

4.6.2 Stockpile; waste dump; spoilbank A deposit at the surface of the mine of mined material, e.g. coal (stockpile), dirt, preparation rejects (waste dump, spoilbank) with a view to evening out fluctuations in fuel supply and demand in the longer term, or to long term dumping.

4.6.3 Dump; spoilbank; tip (opencast, etc., mining) The systematic deposit of overburden, in general in worked out opencast mines or parts of them.

4.6.4 Bunker; bin A holder for storing the products and by-products of mining.

4.6.5 Ditch bunker (opencast, etc., mining) A method of stocking coal to allow for transhipment, in which the sides of the stockpile are formed, in whole or in part, by the natural angle of repose.

4.6 Stockage

4.6.1 Parc à charbon Dépôt au jour des produits valorisables extraits de la mine, réalisé en vue d'assurer la modulation des quantités et l'égalisation des caractéristiques et également destiné à servir de réserve et de compensation des différences entre quantités produites et consommées, qui est *utilisé en permanence,* y compris les dispositifs permettant les transports à l'arrivée et au départ.

4.6.2 Terril (de déchets ou de charbon) Dépôt au jour des produits provenant de la mine, schistes, résidus de la préparation, etc . . . ayant pour objet l'élimination ou la compensation en quantité, *à long terme,* entre l'offre et la demande.

4.6.3 Remblai (exploration à ciel ouvert) Dépôt de déblais réalisé méthodiquement dans des exploitations à ciel ouvert ou dans des parties de celles-ci.

4.6.4 Trémie, silo Réservoir destiné au stockage de produits fournis par l'exploitation minière et de produits obtenus après préparation.

4.6.5 Silo enterré (exploitation à ciel ouvert) Stock permettant d'assurer le relais entre des moyens de transport, dont les murs sont formés en totalité ou partiellement par des talus naturels.

4.6 Lagerung

4.6.1 Lager *Ständig genutzte* Anschüttung über Tage von bergbaulich gewonnenen *verwertbaren* Produkten zur Mengen- und/oder Eigenschaftsvergleichmässigung sowie zur Bevorratung und zum Mengenausgleich zwischen Erzeugung und Verbrauch einschliesslich der Einrichtungen für die Zuführung und Abförderung.

4.6.2 Halde Anschüttung von *bergbaulich anfallenden* Produkten, Bergen, Rückständen der Aufbereitung usw. an der Tagesoberfläche zum Zwecke der Beseitigung oder des *langfristigen* Mengenausgleiches zwischen Angebot und Nachfrage.

4.6.3 Kippe (Tagebau) Planmässige Ablagerung von Abraum in ausgekohlten Tagebauen oder Teilen davon.

4.6.4 Bunker Behälter zur Speicherung von bergbaulich anfallenden Produkten und Verarbeitungsprodukten.

4.6.5 Grabenbunker (Tagebau) Speicher zur Gewährleistung des Umschlags zwischen Transportmitteln, dessen Wandungen ganz oder teilweise von der natürlichen Böschung gebildet werden können.

4.6 Almacenamiento

4.6.1 Parque de carbones Depósito a cielo abierto de los productos útiles extraídos de la mina, destinado a homogeneizar sus características, sirviendo al mismo tiempo de reserva para compensar las diferencias entre las producciones y consumos; es de uso continuo e incluye los dispositivos que facilitan los transportes en descarga y carga.

4.6.2 Pilas de carbón, escombreras Depósito a cielo abierto de los productos salidos de la mina, carbón, estéril, o residuos del lavadero, bien sea como depósito definitivo, en caso de estériles, o para compensar a largo plazo las diferencias entre la oferta y la demanda, en el caso de carbón.

4.6.3 Vertido de escombros, escombreras (cielo abierto) Depósito sistemático del estéril de recubierto realizado en general sobre las zonas a cielo abierto, ya deshulladas o en alguna de sus partes.

4.6.4 Tolva, silo Es un depósito destinado al almacenamiento de los productos de la explotación minera y de los productos ya preparados.

4.6.5 Tolva enterrada (explotación a cielo abierto) Almacenaje del carbón intermedio entre diferentes medios de transporte, en los que los laterales de la pila están formados total o parcialmente por taludes naturales.

Section 5

Extraction and Refining of Liquid Fuels
Extraction et raffinage des combustibles liquides
Gewinnung and Verarbeitung flüssiger Brennstoffe
Extraccion y Refino de los Combustibles Liquidos

5

Extraction and Refining of Liquid Fuels

5.1 Sources and Forms of Liquid Fuels

5.1.1 Oil reservoir; oil deposits A porous, permeable, underground formation containing an individual and separate natural accumulation of producible oil or gas.

5.1.2 Reservoir rock Porous and permeable rock containing producible oil and/or gas in its pore spaces.

5.1.3 Bituminous shale; oil shale; bituminous schist (USA) Bituminous rock that releases hydrocarbons on distillation.

5.1.4 Tar sands; oil sands (Can.) "Tar sands" are a geological phenomenon found especially in Canada, in which bitumen-saturated sands are found near or at the surface of the ground; the hydrocarbon can be separated from the sand by mechanical or thermal processes.

Heavy oil sands are similar to tar sands. Outside Canada the term "oil sands" generally refers to buried sandstone acting as an oil reservoir.

5.1.5 Crude oil reserves The amount of crude oil or gas expected to be recovered profitably from the reservoir rock.

5.1.6 Proved reserves Clearly proven reserves of crude oil extractable under current operating and economic conditions.

5.1.7 Probable reserves Known productive reserves in existing fields expected to respond to improved recovery techniques (installed but not yet fully evaluated or not yet installed but predictable from similar installations).

5.1.8 Possible reserves Reserves from those areas whose oil bearing has not yet been established by production tests, but for which data collected to date favour the presence and extractability of oil.

5.1.9 Bore hole; drill hole; well A hole or well drilled or bored principally by mechanical means in order to explore geological conditions and/or to tap oil deposits.

5.1.10 Crude oil; petroleum Naturally occurring mineral oil consisting essentially of many types of hydrocarbons. Crude oil may be of paraffinic, asphaltic or mixed base, according to the presence of paraffin wax, bitumen or both paraffin wax and bitumen in the residue after atmospheric distillation. In modern technical usage the term petroleum includes gaseous and solid as well as liquid hydrocarbons.

5.1.11 Pretreated crude oil Degassed and dewatered crude oil.

5.1.12 Paraffin base crude; paraffinic crude Petroleum that contains hydrocarbons predominantly of the paraffin series.

5.1.13 Naphthene base crude; naphthenic crude Petroleum that contains hydrocarbons predominantly of the cycloparaffin series.

5.1.14 Mixed base crude Petroleum in which neither the hydrocarbons of the paraffin series nor those of the cycloparaffin series predominate, their proportions being approximately equal.

Extraction et raffinage des combustibles liquides

5.1 Sources et Types de Pétrole

5.1.1 Gisements de pétrole Roche-magasin contenant des réserves de pétrole exploitable.

5.1.2 Roche-magasin Roche poreuse et perméable dans laquelle le pétrole s'est concentré.

5.1.3 Schistes bitumineux Roches bitumineuses qui fournissent des hydrocarbures par distillation à basse température

5.1.4 Sables asphaltiques Sables contenant du pétrole qui peut en être séparé par des traitements mécaniques ou thermiques.

5.1.5 Réserves de pétrole Quantité économiquement exploitable contenue dans la roche-magasin.

5.1.6 Réserves certaines (prouvées) Réserves de pétrole économiquement exploitables dont la présence, a été clairement démontrée.

5.1.7 Réserves probables Réserves dont la présence dans une zone déterminée est démontrée et dont la possibilité d'exploitation est nettement reconnue, mais pour lesquelles les résultats de production font défaut ou sont insuffisants.

5.1.8 Réserves possibles Réserves situées dans des zones pour lesquelles la possibilité d'extraction du pétrole n'est pas confirmée par des essais de production, mais dont les caractéristiques connues indiquent la présence de pétrole et la possibilité d'extraction.

5.1.9 Forage Trous de forage réalisés par des méthodes principalement mécaniques, en vue du repérage des conditions géologiques et/ou de la reconnaissance des gisements de pétrole et/ou de la mise en exploitation.

5.1.10 Pétrole Mélange, en proportions variables d'hydrocarbures solides, liquides ou gazeux dans les conditions normales, qui se présente à l'état naturel, sous pression et température plus ou moins élevées, dans les gisements.

5.1.11 Brut Pétrole extrait du gisement après séparation du gaz et de l'eau éventuellement entraînés.

5.1.12 Pétrole paraffinique, de base paraffinique Pétrole contenant une fraction prédominante d'hydrocarbures de la série paraffinique.

5.1.13 Pétrole naphténique, de base naphténique Pétrole contenant une fraction prédominante d'hydrocarbures de la série cycloparaffinique.

5.1.14 Pétrole mixte, à base mixte Pétrole dans lequel ni les hydrocarbures de la série paraffinique, ni ceux de la série cycloparaffinique, ne sont prédominants, mais dont les proportions sont approximativement égales.

Gewinnung and Verarbeitung flüssiger Brennstoffe

5.1 Quellen und Erdölarten

5.1.1 Erdöllagerstätte Trägergestein mit abbaubaren Erdölvorräten

5.1.2 Trägergestein Poröser und durchlässiger Gesteinskörper, in dem eine Erdölanreicherung vorhanden ist.

5.1.3 Oelschiefer Bituminose Gesteine, die bei der Schwelung Kohlenwasserstoffe abgeben.

5.1.4 Oelsande Erdölhaltige Sande, von denen das Erdöl durch mechanische oder thermische Verfahren getrennt werden kann.

5.1.5 Erdölvorräte Wirtschaftlich abbaubare Erdölmenge im Trägergestein

5.1.6 Sichere Vorräte Eindeutig nachgewiesene und wirtschaftlich förderbare Erdölvorräte.

5.1.7 Wahrscheinliche Vorräte Vorräte, in Gebieten nachgewiesen, die sich an eindeutig produktiv erkannte anschliessen, wo aber Produktionsergebnisse fehlen oder unzureichend sind.

5.1.8 Mögliche Vorräte Vorräte aus jenen Gebieten, deren Erdölführung *nicht* durch Produktionsversuche sichergestellt ist, für deren Vorhandensein und Förderbarkeit jedoch die bisher bekannten Daten sprechen.

5.1.9 Bohrung Mit Hilfe von vorwiegend mechanischen und/oder hydraulischen Methoden abgeteufte Bohrlöcher zur Erkundung der geologischen Verhältnisse und/oder zur Aufschliessung von Erdölvorräten.

5.1.10 Erdöl Gemisch von vorwiegend festen, flüssigen oder gasförmigen Kohlenwasserstoffen unterschiedlicher Zusammensetzung.

5.1.11 Rohöl Gewonnenes Erdöl nach Entgasung und Abscheidung eventuell mitgerissenen Wassers.

5.1.12 Paraffinbasisches Erdöl Erdöl mit überwiegenden Anteilen an Kohlenwasserstoffen der Paraffinreihe. (Im allgemeinen grösser als 75%).

5.1.13 Naphtenbasisches Erdöl Erdöle mit überwiegenden Anteilen an Kohlenwasserstoffen der Cycloparaffinreihe (Naphtene). (Im allgemeinen grösser als 70%).

5.1.14 Gemischt-basisches Erdöl Bei dem weder die Kohlenwasserstoffe der Paraffinreihe noch die der Cycloparaffinreihe überwiegen, sondern deren Anteile annähernd gleich sind.

Extraccion y Refino de los Combustibles Liquidos

5.1 Fuentes y Tipos de Petróleo

5.1.1 Yacimiento, depósito pretolífero Roca almacén que contiene reservas de petróleo explotables.

5.1.2 Roca almaćen Roca permeable y porosa en la que se ha concentrado el petróleo.

5.1.3 Esquistos o pizarra bituminosos Roca bituminosa que desprende hidrocarburos al ser destilada a baja temperatura.

5.1.4 Arenas asfálticas Arenas que contienen petróleo que puede ser separado por procedimientos mecánicos o térmicos.

5.1.5 Reservas de petróleo Cantidad de petróleo económicamente explotable contenida en una roca almacén.

5.1.6 Reservas seguras (Probadas) Reservas de petróleo explotables económicamente cuya presencia está claramente demostrada.

5.1.7 Reservas probables Reservas cuya presencia en una zona determinada está demostrada y cuya posibilidad de explotación está claramente reconocida, pero de las que no se dispone de resultados de producción o bien se dispone de resultados insuficientes en esa zona.

5.1.8 Reservas posibles Reservas situadas en zonas para las que no se ha confirmado, por medio de ensayos de producción, la posibilidad de extracción de petróleo, pero cuyas características conocidas indican la presencia de petróleo y la posibilidad de su extracción.

5.1.9 Sondeo, perforación Perforaciones realizadas por medios mecánicos principalmente, a fin de establecer puntos de referencia de las condiciones geológicas y/o del reconocimiento de los yacimientos de petróleo y/o de la puesta en explotación.

5.1.10 Petróleo Mezcla en proporciones variables, de hidrocarburos sólidos liquidos o gaseosos, en condiciones normales, y que se presenta en estado normal, sometido a presión y temperatura más o menos elevadas, en los yacimientos.

5.1.11 Crudo Petróleo extraido del yacimiento, una vez separado del gas y del agua que pueda haber arrastrado eventualmente.

5.1.12 Petróleo parafínico, de base parafínica Petróleo que contiene una base pred minante de hidrocarburos de la serie parafínica.

5.1.13 Petróleo nafténico, de base nafténica Petróleo que contiene una base pred minante de hidrocarburos de là serie cicloparafínica.

5.1.14 Petróleo mixto, de base mixta Petróleo en el que no predominan los hidrocarburos de la serie parafínica ni de la cicloparafínica pero cuyas proporciones son aproximadamente iguales.

5

5.2 Refining

5.2.1 Pretreatment The preparation of crude oil by degasifying it, removing water, salt, etc., from it, in order to be able to use it as a feedstock for primary distillation.

5.2.2 Separation processes Processes for the splitting up of hydrocarbon mixtures on the basis of their chemical and physical properties.

5.2.3 Distillation The separation of mixtures of hydrocarbons into several fractions by vaporization followed by condensation. Heating of the feedstock is generally carried out in pipe stills and fractionation in columns; the distillation may be either atmospheric distillation or vacuum distillation according to the end products required.

5.2.4 Conversion processes Processes in which the principal reaction involves the carbon bonds.

5.2.5 Treating/finishing processes Processes in which reactions take place affecting non-hydrocarbon substances—heterogeneous compounds—in oil fractions, e.g. sweetening, caustic washing, desulphurisation.

5.2.6 Cracking Increasing the relative proportions of lighter or more volatile components of an oil by breaking down the larger hydrocarbon molecules into smaller molecules.

5.2.7 Reforming A process (cyclization, aromatization, etc.) whereby light petroleum fractions are treated to yield gasoline having essentially a higher aromatics content and hence a higher octane rating than the feedstock.

5.2.8 Isomerization In the petroleum industry the process is employed for converting straight-chain or slightly branched-chain paraffins into heavily branched-chain paraffins.

5.2.9 Molecular sieve process A process in which separation is effected by means of a molecular sieve, which enables molecules within similar boiling ranges to be separated by virtue of their geometric characteristics.

5.3 Final Products; Finished Products

5.3.1 Liquefied petroleum gas (LPG); liquefied refinery gas (LRG) (USA) A mixture of light hydrocarbons, gaseous under conditions of normal temperature and pressure and maintained in the liquid state by increase of pressure or lowering of temperature; the principal components are propane, propene, butanes and butenes. Composition and properties depend on national specifications.

5.3.2 Gasoline; petrol; motor spirit; motor gasoline (USA) Refined petroleum distillate, normally boiling within the limits of 30 to 220°C, which, combined with certain additives, is used as fuel for spark-ignition engines. By extension, the term is also applied to other products boiling within this range. Composition and properties depend on national specifications.

Alternative definition (USA): Fuel for internal combustion engines with spark ignition comprising refinery products within the gasoline range to be marketed as motor gasoline without further processing (blending excepted).

5.3.3 Diesel oil; diesel fuel; derv fuel (UK); **auto diesel oil** A liquid hydrocarbon mixture in the gas oil range for use in compression-ignition internal combustion engines. Composition and properties depend on national specifications.

5.2 Raffinage

5.2.1 Traitements préliminaires Épuration du pétrole par dégazage, séparation de l'eau, déminéralisation, etc. . . avant son envoi à la distillation primaire.

5.2.2 Séparation Décomposition de mélanges d'hydrocarbures par des procédés physiques et chimiques.

5.2.3 Distillation Séparation de mélanges d'hydrocarbures en plusieurs fractions, par vaporisation suivie de condensation. Le chauffage des produits mis en oeuvre s'effectue en général dans des fours tubulaires et la séparation dans des colonnes : selon la nature des produits finis, on met en oeuvre une distillation sous pression atmosphérique ou une distillation sous vide.

5.2.4 Procédés de conversion Ensemble des traitements chimiques où la réaction principale s'effectue sur les liaisons du carbone.

5.2.5 Procédés d'épuration Ensemble de traitements où les réactions s'effectuent sur les corps autres que des hydrocarbures contenus dans des fractions pétrolières, par exemple : désacidification, lavage alcalin, désulfuration.

5.2.6 Craquage Transformation par rupture des grosses molécules d'hydrocarbures pour en obtenir de plus petites en vue d'augmenter la proportion des produits plus légers et plus volatiles.

5.2.7 Reformage Transformation par cyclisation, aromatisation, etc. d'hydrocarbures légers avec séparation d'hydrogène en vue d'augmenter. le contenu aromatique et l'indice d'octane.

5.2.8 Isomérisation Transformation de paraffines à chaîne droite ou faiblement ramifiées en paraffines à chaînes fortement ramifiées.

5.2.9 Procédé de séparation par tamis moléculaire Séparation de molécules ayant des zones d'ébullition similaires suivant leur caractéristiques géométriques.

5.3 Produits finis

5.3.1 Gaz de pétrole liquéfiés Hydrocarbures en $C_3 - C_4$ et leurs mélanges, spécifiés par leurs caractéristiques.

5.3.2 Essence auto, essence de tourisme Mélanges d'hydrocarbures liquides spécifiés par leurs caractéristiques et destinés aux moteurs à explosion à allumage commandé.

5.3.3 Carburant Diesel, gaz-oil moteur Mélanges d'hydrocarbures liquides, spécifiés par leurs caractéristiques et destinés aux moteurs à explosion à auto-allumage.

5.2 Verarbeitung

5.2.1 Vorbehandlung Aufbereitung des Erdöls durch Entgasung, Wasserabscheidung, Entsaizung udgl., um es als Einsatz für die Primärdestillation verwenden zu können.

5.2.2 Trennung Zerlegung von Kohlenwasserstoffgemischen nach chemischem und physikalischem Verhalten.

5.2.3 Destillation Kontinuierliche Trennung von Kohlenwasserstoffgemischen in mehrere Fraktionen durch Verdampfung und anschliessendes Kondensieren. Die Erwärmung des Einsatzproduktes erfolgt im allgemeinen in Röhrenöfen, die Trennung in Destillationskolonnen; je nach Art des Fertigproduktes wendet man die atmosphärische Destillation oder die Vakuumdestillation an.

5.2.4 Umwandlungsverfahren Eine Gruppe von Verfahren, bei welchen die Hauptreaktion an den Kohlenstoffbindungen stattfindet.

5.2.5 Veredlung Verfahren, bei denen Reaktionen an den Nichtkohlenwasserstoffen - Heteroverbindungen - von Erdölfraktionen stattfinden, z.B. Süssen, Laugenwäsche, Entschwefelung.

5.2.6 Spalten Abbau von grösseren Kohlenwasserstoffmolekülen zu kleineren, zur Erhöhung des Anteils an leichteren und flüchtigeren Produkten.

5.2.7 Reformieren Umwandlung (Cyklisierung, Aromatisierung usw.) von leichten Kohlenwasserstoffen unter Abspaltung von Wasserstoff, zur Erhöhung des aromatischen Anteils und der Oktanzahl.

5.2.8 Isomerisation Umwandlung von kettenförmigen oder schwach verzweigten Paraffinen zu solchen mit höherem Verzweigungsgrad.

5.2.9 Molekularsiebverfahren Trennung nach geometrischen Eigenschaften der Moleküle ähnlichen Siedebereiches.

5.3 Fertigprodukte

5.3.1 Flüssiggas C_3 - C_4 - Kohlenwasserstoffe und deren Gemische, spezifiziert durch Kennwerte.

5.3.2 Ottokraftstoff Flüssige Kohlenwasserstoffgemische zur Verwendung in Verbrennungskraftmaschinen mit Fremdzündung. Spezifikation durch Kennwerte.

5.3.3 Dieselkraftstoff Flüssige Kohlenwasserstoffgemische zur Verwendung in Verbrennungskraftmaschinen mit Selbstzündung. Spezifikation durch Kennwerte.

5.2 Refino

5.2.1 Pretratamiento Preparación del crudo de petróleo, extrayéndole el gas, agua, minerales etc. antes de su envío a la destilación primaria.

5.2.2 Separación Descomposición de las mezclas de hidrocarburos por procedimientos físicos y químicos.

5.2.3 Destilación Separación de mezclas de hidrocarburos en varias fracciones, por vaporización seguida de condensación. El calentamiento de los productos a tratar se realiza, por lo general, en hornos tubulares y la separación en columnas; según la naturaleza de los productos finales, se efectúa mediante una destilación a la presión atmosférica o mediante una destilación en vacío.

5.2.4 Procedimientos de conversión Conjunto de operaciones químicas en las que la reacción principal implica enlaces del carbono.

5.2.5 Refino quimico Conjunto de tratamientos en los que las reacciones tienen lugar alterando las sustancias que no son hidrocarburos compuestos heterogéneos - en fracciones de petróleo; v.g. desacidificación, lavado alcalino, desulfurización.

5.2.6 Craqueo Transformación, por ruptura de las grandes moléculas, de hidrocarburos para obtenerlas más pequeñas a fin de aumentar la proporción de productos más ligeros y volátiles.

5.2.7 Reformado Transformación por ciclización, aromatización etc. de hidrocarburos ligeros con separación de hidrógeno, a fin de aumentar el contenido aromático y el índice de octano.

5.2.8 Isomerizacion Transformación de parafinas de cadena lineal o débilmente ramificadas en parafinas de cadena muy ramificada.

5.2.9 Separacion por cribado molecular Separación de moléculas que tienen zonas de ebullición semejantes, de acuerdo con sus características geométricas.

5.3 Productos Acabados

5.3.1 Gases licuados de petróleo Hidrocarburos en C_3 - C_4 y sus mezclas, especificados por sus características.

5.3.2 Gasolina auto Mezcla de hidrocarburos líquidos, que se especifican según sus características y destinados a motores de explosión, sin autoencendido.

5.3.3 Combustible diesel Mezcla de hidrocarburos líquidos, que se especifican según sus características y destinado a los motores de explosión sin autoencendido.

5

5.3.4 Jet fuel; aviation turbine fuel (ATF) (UK) Petroleum distillates used as a source of energy in systems of jet propulsion. By extension, fuel suitable for use in aircraft gas turbines. Composition and properties depend upon national specifications.

5.3.5 Fuel oils Hydrocarbon mixtures (liquid or liquefiable petroleum products) normally without light boiling fractions, for use in burners. Composition and properties depend on national specifications.

5.4 Storage and Transport

5.4.1 Storage An installation comprising one storage tank or two or more storage tanks (a tank farm) for storing crude oil and/or petroleum products.

5.4.2 Tanker; tank vessel (USA); **tank ship** (USA) A merchant ship designed for the transport of liquid cargoes.

5.4.3 Rail tanker; rail tank car (Can.); **tanker** (USA) A railway freight car designed for the transport of liquids.

5.4.4 Road tanker; road tank car; tank truck (USA) A road vehicle designed for the transport of liquids.

5.4.5 Pipeline A continuous pipe conduit above or below ground complete with facilities for transporting fluids, generally including compressor stations at regular intervals along its length.

5.3.4 Carburants Caractéristiques selon spécifications.

5.3.5 Mazout, fuel Mélanges d'hydrocarbures, spécifiés par leurs caractéristiques, mais ne comprenant pas de fraction à bas point d'ébullition et destinés aux brûleurs d'installation thermiques.

5.4 Stockage et Transport

5.4.1 Stockage Installation comportant un ou plusieurs réservoirs en vue du stockage de combustibles liquides.

5.4.2 Bateau-citerne, pétrolier Cargo destiné au transport de chargements liquides.

5.4.3 Wagon-citerne Wagon de chemin de fer destiné au transport de chargements liquides.

5.4.4 Camion-citerne, citerne routière Véhicule routier destiné au transport de chargements liquides.

5.4.5 Conduite de transport, oléoduc Canalisations généralement souterraines, à côté desquelles il existe le plus souvent des stations de pompage intermédiaires.

5.3.4 Düsenkraftstoff Spezifikation durch Kennwerte.

5.3.4 Carburantes Características de acuerdo con las especificaciones correspondientes.

5.3.5 Heizöle Kohlenwasserstoffgemische ohne Anteil leicht siedender Fraktionen zur Verwendung für Brenner.
Spezifikation durch Kennwerte.

5.3.5 Fuel combustible Mezclas de hidrocarburos, especificados según sus características pero que no incluyen fracciones de bajo punto de ebullición y que están destinados a los quemadores de las instalaciones térmicas.

5.4 Lagerung und Transport

5.4.1 Lager Einrichtung, die aus einem oder mehreren Behältern für die Einlagerung von flüssigen Brennstoffen besteht.

5.4.2 Tankschiffe Frachtschiff zum Transport flussiger Ladung, z.B. Flüssigerdgas.

5.4.3 Kesselwagen Kesselwagen ist ein Eisenbahngüterwagen zum Transport von Flüsigkeiten, z.B. von Flüssiggasen.

5.4.4 Tankwagen Tankwagen ist ein Strassenfahrzeug zum Transport flüssiger Ladung, z.B. Flüsiggas.

5.4.5 Pipeline Im wesentlichen unterirdisch verlegte Rohrleitungen, bei denen meist in bestimmten Abständen Pumpstationen zwischengeschaltet sind.

5.4 Almacenamiento y transporte

5.4.1 Almacenamiento Instalación que cuenta con uno o varios depósitos con la finalidad de acopiar los combustibles líquidos.

5.4.2 Buque cisterna, petrolero Cargo destinado al transporte de cargamentos líquidos.

5.4.3 Vagón-cisterna Vagón de ferrocarril destinado al transporte de cargamentos líquidos.

5.4.4 Camión cisterna Vehículo para el transporte, por carretera, de cargamentos líquidos.

5.4.5 Oleoducto Canalizaciones, generalmente subterráneas, a cuyo servicio se dispone, frecuentemente, de instalaciones intermedias de bombeo.

5

Section 6

Gas Industry
Industrie gazière
Begriffe der Gaswirtschaft
Industria del Gas

6

Gas Industry

6.1 General Terms

6.1.1 Flow; rate of flow; output; throughput; flow rate The quantity of gas flowing through a pipe in unit time.

6.1.2 Load curve; output curve A graph showing load, output etc. against time.

6.1.3 Base load That part of total demand that does not vary over a given period (day, month, year).

6.1.4 Coincidence factor The ratio, expressed as a numerical value or as a percentage, of the simultaneous maximum demand of a group of gas appliances or consumers within a specified period, to the sum of their individual maximum demands within the same period. *Diversity factor* is the reciprocal of the coincidence factor.

6.1.5 Gas volume The space occupied by a specified mass of gas under specified conditions.

6.1.6 Metric standard conditions; standard conditions; standard reference conditions The conditions to which the gas volume or other properties of the gas are referred. "Metric standard conditions" are 1.01325 bar (101,325 Pa), 15°C, dry; "standard conditions" may be in terms of metric or other units, dry or saturated, as specified in the national standards of the country concerned; "standard reference conditions" are 30 in. Hg, 60°F, saturated.

6.1.7 Quantity of gas The gas measured as a volume under specified conditions.

6.1.8 Gas consumption The metered quantity of gas intended for consumption in gas appliances.

6.1.9 Wholesale gas purchase The quantity of gas purchased by one gas supply undertaking from another.

6.1.10 Gas demand The amount of gas required in unit time by a consumer or at a specific point (e.g. local point of supply, appliance). Gas demand in the context of load forecasting would be termed *anticipated demand.*

6.1.11 Distribution; send out (USA); **output** (USA) Gas delivered to a delivery or distribution system from a specific point (e.g. plant or metering station).
Note In some countries these terms have more specific meanings attached to them than indicated above.

6.1.12 Gas purchase The quantity of gas received by a purchaser at his supply point.

6.1.13 Unaccounted-for gas; leak rate; leakage rate "Unaccounted-for gas" is the difference between send-out and gas sold or otherwise usefully accounted for. "Leak rate" and "leakage rate" require to be stated in m³/h or analogous units.

6.1.14 Relative density The ratio of the weight of unit volume of dry gas to that of unit volume of dry air under the same conditions of temperature and pressure. Also called *specific gravity.*

Industrie gazière

6.1 Termes généraux

6.1.1 Débit Quantité de gaz qui s'écoule dans une canalisation de gaz pendant l'unité de temps.

6.1.2 Courbe de débit (Courbe des émissions) Représentation graphique de l'évolution du débit dans le temps.

6.1.3 Charge de base Partie constante du débit total considéré pendant une période donnée (par exemple : jour, mois, année).

6.1.4 Coefficient de simultanéité de foisonnement "Quotient de la pointe de la demande globale simultanée par la somme des *pointes* des demandes individuelles" ; Le "facteur de diversité" est l'inverse du coefficient de simultanéité".

6.1.5 Volume du gaz Espace occupé par une masse déterminée de gaz dans des conditions déterminées.

6.1.6 Conditions normales Volume et autres propriétés physiques du gaz rapportés à une pression de 1,01325 bar (101 325Pa) et à une température de 15°C.
N.B.: La température de 15°C n'est pas utilisée dans tous les pays.

6.1.7 Quantité de gaz Quantité de matière sous forme gazeuse mesurée en volume dans des conditions données.

6.1.8 Consommation de gaz Quantité de gaz mesurée en vue de la consommation par des appareils d'utilisation.

6.1.9 Livraison de gaz Quantité de gaz qu'une entreprise de distribution de gaz reçoit d'une autre entreprise.

6.1.10 Demande de gaz Quantité de gaz, déterminée par estimation, qui sera prélevée par un consommateur ou en un point déterminé (par exemple : poste de livraison).

6.1.11 Emission (de gaz) Quantité de gaz émise en un point déterminé du réseau qui doit être désigné avec précision.

6.1.12 Enlèvement (de gaz) Quantité de gaz prélevée par un consommateur à son point de livraison.

6.1.13 Pertes de gaz Différence entre l'émission à partir de l'usine de production de gaz et la totalité des fournitures utilisables.

6.1.14 Densité par rapport à l'air Quotient de la densité du gaz par celle de l'air, considérées dans les mêmes conditions.

Begriffe der Gaswirtschaft

6.1 Allgemeine Begriffe

6.1.1 **Belastung** Belastung ist die durch eine Rohrleitung in der Zeiteinheit fliessende Gasmenge.

6.1.2 **Ganglinie** Ganglinie ist die graphische Darstellung des zeitlichen Verlaufs der Belastung.

6.1.3 **Grundlast** Grundlast ist der während einer Zeitspanne (z.B. Tag, Monat, Jahr) gleichbleibende Teil der Gesamtbelastung.

6.1.4 **Gleichzeitigkeitsfaktor** Gleichzeitigkeitsfaktor ist das Verhältnis zwischen der tatsächlich auftretenden Verbrauchsspitze und der Summe der möglichen Verbrauchsspitzen der einzelnen Abnehmer.
Der Diversitätsfaktor ist die Umkehrung des Gleichzeitigkeitsfaktor.

6.1.5 **Gasvolumen** Gasvolumen ist der von einer bestimmten Masse des Gases unter bestimmten Zustandsbedingungen eingenommene Raum.

6.1.6 **Standardzustand** Standardzustand ist ein Druck von 1,01325 bar (101 325 Pa) und eine Temperatur von 15°C, worauf trockene Gasvolumina und andere physikalische Eigenschaften bezogen werden.
Nota : Die Temperatur von 15°C ist nicht in allen Ländern gebräuchlich.

6.1.7 **Gasmenge** Gasmenge ist die unter gegebenen Zustandsbedingungen als Volumen gemessene stoffliche Substanz des Gases.

6.1.8 **Gasverbrauch** Gasverbrauch ist die gemessene Gasmenge zur Verwendung in Gasverbrauchsgeräten.

6.1.9 **Gasbezug** Gasbezug ist die jenige Gasmenge, die ein Gasversorgungsunternehmen von einem anderen bezieht.

6.1.10 **Gasbedarf** Gasbedarf ist die beim Abnehmer oder einer bestimmten Stelle (z.B. Uebergabestation) zu erwartende Gasabnahme.

6.1.11 **Gasabgabe** Gasabgabe ist die an einer bestimmten, näher zu bezeichnenden Stelle eines Gasnetzes abgegebenen Gasmenge.

6.1.12 **Gasabnahme** Gasabnnahme ist die von einem Abnehmer an seiner Uebergangstelle entnommene Gasmenge.

6.1.13 **Gasverlust** Gasverlust ist der Unterschied zwischen Abgabe ab Gaserzeugungsbetrieb und der gesamten nutzbaren Abgabe.

6.1.14 **Dichteverhältnis** Dichteverhältnis ist der Quotient aus der Dichte des Gases und der der Luft unter gleichen Zustandsbedingungen.

Industria del Gas

6.1 Conceptos Generales

6.1.1 **Caudal** Cantidad de gas que fluye a través de una tubería en la unidad de tiempo.

6.1.2 **Curva de caudal** Representación gráfica de la variación del caudal a lo largo del tiempo.

6.1.3 **Carga de base** Parte constante del caudal total considerado durante un período de tiempo dado (por ej. día, mes, año).

6.1.4 **Coeficiente de simultaneidad** Cociente entre la punta de demanda global simultánea y la suma de las puntas de las demandas individuales. El "factor de diversidad" es la inversa del coeficiente de simultaneidad.

6.1.5 **Volumen del gas** Espacio ocupado por una masa determinada de gas en determinadas condiciones.

6.1.6 **Condiciones normales** Volumen y otras propiedades físicas del gas seco medidos a 1,01325 bar (101 325 Pa) y 15°C.
Nota. La temperatura de 15° C no se utiliza en todos los países.

6.1.7 **Cantidad de gas** Cantidad de materia en forma gaseosa medida en volumen en las condiciones dadas.

6.1.8 **Consumo de gas** Cantidad de gas medida para ser consumida en los aparatos de utilización.

6.1.9 **Suministro de gas al por mayor** Cantidad de gas que una empresa de distribución de gas recibe de otra empresa.

6.1.10 **Demanda de gas** Cantidad de gas, que se determina por estimación, que será retirada por un consumidor en un punto determinado (por ejemplo: un puesto de suministro).

6.1.11 **Suministro (de gas)** Cantidad de gas suministrada en un punto determinado de la red, punto que debe señalarse con precisión.

6.1.12 **Gas adquirido** Es la cantidad de gas recibida por un comprador en el punto de suministro.

6.1.13 **Pérdidas de gas** Diferencia entre el suministro a la salida de la fábrica y el total de las ventas utilizables.

6.1.14 **Densidad relativa con relación al aire** Relación entre la densidad del gas y la del aire, consideradas en las mismas condiciones.

6

6.1.15 Gross calorific value; gross heating value The amount of heat liberated by complete combustion, under specified conditions, of unit volume of a gas, the *water* produced by the combustion of the gas being assumed to be *completely condensed* and its latent heat released, the other products of combustion being referred to the standardised test conditions as applied in different countries. The specified conditions are generally, in the UK and USA, 30 in. Hg and 60°F and, in countries using the metric system or SI system, 1.01325 bar or 101,325 Pa and 0°C. In the UK and USA calorific value is normally expressed in Btu per cu ft; in countries using the metric or SI system, in kcal per Nm³ or kJ per Nm³.

6.1.16 Net calorific value; net heating value The amount of heat liberated by the complete combustion, under specified conditions, of unit volume of a gas, the *water* produced by the combustion of the gas being assumed to remain as a *vapour,* the other products of combustion being referred to the standardised test conditions as applied in different countries. Hence the net calorific value is the gross calorific value less the latent heat of evaporation of the water that formed during combustion of the fuel. See also under 6.1.15 above.

6.1.17 Wobbe Index; Wobbe number The ratio of the gross (or net) calorific value of the gas to the square root of the relative density of the gas. It presents a measure of the heat release when a gas is burned at constant gas supply pressure. The heat release is then directly proportional to the orifice area and the Wobbe number.

6.1.18 Water vapour dewpoint The temperature at which at a given pressure water vapour in the gas condenses.

6.1.19 Hydrocarbon dewpoint The temperature at which at a given pressure hydrocarbon vapours in the gas condense.

6.1.20 Combustion velocity; burning velocity A physical characteristic of a gas; it is the maximum velocity, relative to the unburned gas, with which a plane, one-dimensional flame front travels along the normal to its surface.

6.1.21 Flammability limits; explosion limits; explosive limits The upper and lower limits of the concentration of a combustible gas in air or oxygen, between which the mixture is explosive or flammable; such limits may vary according to prevailing temperature and pressure.

6.1.22 Pressure rating; rated pressure The design pressure of gas equipment.

6.1.23 Working pressure The gas pressure at which the equipment is operated.

6.1.24 Gas pressure The pressure of the gas above atmospheric pressure, more accurately termed gauge pressure.

6.1.25 Test pressure The pressure to which equipment is tested.

6.1.26 Pressure range The subdivision of the working pressure into low, medium and high pressure ranges.

6.1.27 Rated heat input; heat input rating The quantity of potential heat calculated on the basis of calorific value (net or gross according to specification) supplied to the burner in unit time that, according to the manufacturer's name plate, may not be exceeded.

6.1.28 Rated useful heat; rated useful heat output The quantity of heat liberated by the burner in unit time when operating at its rated heat input.

6.1.15 Pouvoir calorifique supérieur (p.c.s.) Quantité de chaleur libérée par la combustion complète d'un mètre cube de gaz (1,01325 bar ou 101325 Pa, 0°C) lorque l'eau formée pendant le combustion se trouve à *l'état liquide* et que les produits de combustion sont évacués dans les conditions d'essais normalisées dans les différents pays.

6.1.16 Pouvoir calorifique inférieur (p.c.i.) Quantité de chaleur libérée lors de la combustion complète d'un mètre cube de gaz (1,01325 bar ou 101325 Pa, 0°C) lorsque l'eau formée pendant le combustion demeure à *l'état de vapeur* et que les produits de combustion sont évacués dans les conditions d'essais normalisées dans les différents pays.

6.1.17 Indice de Wobbe Quotient du pouvoir calorifique inférieur (ou du pouvoir calorifique supérieur) par la racine carrée de la densité du gaz par rapport à l'air.

6.1.18 Point de rosée de l'eau Température à laquelle la vapeur d'eau contenue dans un gaz se condense à une pression donnée.

6.1.19 Point de rosée des hydrocarbures Température à laquelle se condensent les vapeurs d'hydrocarbures contenues dans un gaz à une pression donnée.

6.1.20 Vitesse de déflagration, vitesse de combustion Vitesse de combustion d'un mélange air/gaz ou oxygène/gaz dans des conditions d'écoulement laminaire.

6.1.21 Limites d'inflammabilité Concentrations limites, inférieure et supérieure, d'un gaz combustible dans l'air ou dans l'oxygène entre lesquelles le mélange est inflammable. Ces concentrations varient avec la température et la pression.

6.1.22 Pression nominale Pression pour laquelle une installation de gaz a été réalisée.

6.1.23 Pression de service Pression du gaz sous laquelle fonctionne une installation.

6.1.24 Pression de gaz Excédent de pression du gaz par rapport à la pression atmosphérique.

6.1.25 Pression d'essai Pression à laquelle les installations, appareillages, appareils, tuyauteries, canalisations, accessoires, etc. . . sont essayés.

6.1.26 Domaine de pression Classification de la pression de service en basse pression, moyenne pression et haute pression.

6.1.27 Débit calorifique nominal Débit calorifique indiqué par le fabricant sur la plaque signalétique (rapporté au pouvoir calorifique supérieur ou inférieur) et qui ne doit pas être dépassé lors du réglage. Les débits calorifiques sont généralement donnés pour le pouvoir calorifique supérieur dans l'industrie du gaz.

6.1.28 Puissance nominale, puissance utile Quantité de chaleur rendue effectivement utilisable dans l'unité de temps par un dispositif d'utilisation du gaz au débit calorifique nominal.

6.1.15 Brennwert Brennwert ist die Wärmemenge, die bei vollständiger Verbrennung von 1 m³ Gas (1,01325 bar bzw. 101325 Pa; 0°C) frei wird, wenn das bei der Verbrennung gebildete Wasser *flüssig* vorliegt und die Verbrennungsprodukte auf die in den jeweiligen Ländern genormten Prüfbedingungen zurückgeführt werden.

6.1.16 Heizwert Heizwert ist die Wärmemenge, die bei vollständiger Verbrennung von 1 m³ Gas (1,01325 bar bzw. 101325 Pa; 0°C) frei wird, wenn das bei der Verbrennung gebildete Wasser *dampfförmig* vorliegt und die Verbrennungsprodukte auf die in den jeweiligen Ländern genormten Prüfbedingungen zurückgeführt werden.

6.1.17 Wobbe-Index Wobbe-Index ist der Quotient aus Heizwert (oder Brennwert) und der Quadratwurzel des Dichteverhältnisses des Gases.

6.1.18 Wasserdampftaupunkt Wasserdampftaupunkt ist die Temperatur, bei der im Gas enthaltener Wasserdampf bei gegebenem Druck kondensiert.

6.1.19 Kohlenwasserstofftaupunkt Kohlenwasserstofftaupunkt ist die Temperatur, bei der im Gas enthaltenen Kohlenwasserstoffdämpfe bei gegebenem Druck kondensieren.

6.1.20 Zündgeschwindigkeit Zündgeschwindigkeit ist die Geschwindigkeit der Verbrennung eines Gas-Luft- oder Gas-Sauerstoff-Gemisches unter laminaren Strömungsbedingungen.

6.1.21 Zündgrenzen Zündgrenzen sind die druck- und temperaturabhängige untere und obere Grenzkonzentrationen eines brennbaren Gases in Luft oder Sauerstoff, zwischen denen das Gemisch zündfähig ist.

6.1.22 Nenndruck Nenndruck ist der Druck, für den eine Gasanlage ausgelegt ist.

6.1.23 Betriebsdruck Betriebsdruck ist der Gasdruck, mit dem eine Gasanlage betrieben wird.

6.1.24 Gasdruck Gasdruck ist der Ueberdruck des Gases gegenüber dem atmosphärischen Druck.

6.1.25 Prüfdruck Prüfdruck ist der Druck, mit dem Anlagen, Apparate, Geräte, Rohre, Rohrleitungen, Armaturen usw. geprüft werden.

6.1.26 Druckbereich Druckbereich ist die Unterteilung des Betriebsdruckes in die Bereiche Niederdruck, Mitteldruck und Hochdruck.

6.1.27 Nennwärmebelastung Nennwärmebelastung ist die vom Hersteller auf dem Geräteschild angegebene Wärmebelastung (bezogen auf den Brenn- oder Heizwert), die bei der Einstellung nicht überschritten werden darf.

Nota : In der Gasindustrie wird sie gewöhnlich auf den Brennwert bezogen.

6.1.28 Nennwärmeleistung Nennwärmeleistung ist die bei Nennwärmebelastung von einer Gasverbrauchseinrichtung in der Zeiteinheit tatsächlich nutzbar gemachte Wärmemenge.

6.1.15 Poder calorífico superior (p.c.s.) Cantidad de calor desprendida por la combustión completa de un metro cúbico de gas (1.01325 bar ó 101325 Pa. 0° C) cuando el agua formada durante la combustión se encuentra *en estado liquido* y los productos de la combustión son evacuados en las condiciones de ensayo normalizadas en los diferentes países.

6.1.16 Poder calorífico inferior (p.c.i.) Cantidad de calor desprendida en la combustión completa de un metro cúbico de gas (1.01325 bar 0 101325 Pa. 0° C) cuando el agua formada durante la combustión permanece *en estado de vapor* y los productos de la combustión son evacuados en las condiciones de ensayo normalizadas en los diferentes países.

6.1.17 Indice de Wobbe Cociente entre el poder calorífico inferior (o el poder calorífico superior) del gas y la raiz cuadrada de la densidad relativa del gas con relación al aire.

6.1.18 Punto de rocío del vapor de agua Temperatura a la que, a una presión dada, se condensa el vapor de agua contenido en el gas.

6.1.19 Punto de rocío de un hidrocarburo Temperatura a la que, a una presión dada, se condensan los vapores de hidrocarburo contenido en el gas.

6.1.20 Velocidad de deflagración, velocidad de combustión Velocidad de combustión de una mezcla aire/gas u Oxígeno/gas en condicionee de corriente laminar.

6.1.21 Limites de inflamabilidad Concentraciones límites superior e inferior, de un gas combustible en el aire o en el oxígeno, entre las que la mezcla es inflamable. Estas concentraciones varían con la temperatura y la presión.

6.1.22 Presión nominal Presión de proyecto de una instalación de gas.

6.1.23 Presión de servicio Presión de gas a la que funciona una instalación.

6.1.24 Presión de gas Excedente de la presión del gas con relación a la presión atmosférica.

6.1.25 Presión de prueba Presión a la que se ensayan las instalaciones, equipos aparatos, tuberías, canalizaciones, accesorios etc. . .

6.1.26 Nivel de presión Clasificación de la presión de servicio en baja, media y alta presión.

6.1.27 Caudal calorífico nominal Cantidad de calor en la unidad de tiempo, indicado por el fabricante en la placa de características (referido al poder calorífico superior o inferior) y que no debe rebasarse después del reglaje. En la industria del gas los caudales caloríficos se dan generalmente para el p.c.s.

6.1.28 Poder calorífico normal, poder calorífico útil Cantidad de calor efectivamente utilizable en la unidad de tiempo, por un dispositivo de utilización del gas, circulando el caudal calorifico nominal.

6

6.2 Types of Gas

6.2.1 Fuel gases; gaseous fuels; combustible gases Gases or gas mixtures that burn with air or oxygen and are used mainly for heat generation.

6.2.2 Families of gases Gaseous fuels whose combustion characteristics are in large measure similar, i.e. they have similar Wobbe numbers, which make them interchangeable. The first family of gases comprises town gas (hydrogen-rich gaseous fuels); the second family of gases comprises natural gases, gases associated with petroleum and gases interchangeable with these; and the third family of gases comprises liquefied petroleum gases (propane and butane). The Wobbe numbers of the families of gases are as follows:

Family	Wobbe Number	
	(MJ/standard m³)	(Btu/ft³ at s.r.c.)
1	24.4 - 28.8	600 - 785
2	48.2 - 53.2	1040 - 1450
3	72.6 - 87.8	1940 - 2300

6.2.3 Natural gases Gases, consisting mainly of methane, occurring naturally in underground deposits.

6.2.4 Associated gases; casinghead gases (USA, Can.) Natural gases associated with oil accumulations; they may contain large fractions of higher hydrocarbons. The gases may be dissolved in the oil under the reservoir temperatures and pressures *(solution gas)* or may form a cap of free gas above the oil in the reservoir *(gas cap gas)*. In the USA and Canada the term "associated gas" is applied to free natural gas in immediate contact, but not in solution, with crude oil in the reservoir.

6.2.5 Liquefied petroleum gases (LPG); liquefied refinery gases (LRG) (USA) Mixtures of light hydrocarbons, gaseous under conditions of normal temperature and pressure and maintained in the liquid state by increase of pressure or lowering of temperature. The principal components are propane, propene, butanes and butenes.

6.2.6 Refinery gases Gases produced during the refining and processing of petroleum and petroleum products; they consist mainly of C_1 to C_4 hydrocarbons with variable amounts of free hydrogen, nitrogen and possibly hydrogen sulphide.

6.2.7 Coke-oven gases Gases produced in coke ovens.

6.2.8 High-pressure gasification gases Gaseous fuels produced by reacting solid or liquid fuels with a gasification medium (e.g. oxygen/steam mixture) under high pressure; they may also be gaseous fuels produced by the conversion of liquid fuels by thermal or catalytic processes at high pressure.
Note. In cases in which a gas is the feedstock, a more specific term would be used, e.g. reformed gases.

6.2.9 Cracked gases Gaseous fuels that are produced from liquid or gaseous hydrocarbons by thermal or thermal-catalytic conversion.

6.2.10 Town gas; city gas (USA) Gases manufactured for public supply with a Wobbe number range of 24.4-28.8 MJ/standard m³ (600-785 Btu/ft³ at s.r.c.); they fall within the first family of gases.

6.2.11 Producer gases; lean gases Gaseous fuels produced by continuously gasifying solid fuel in air or in a mixture of air and steam. They are gases of low calorific value and hence referred to as *lean gases.*

6.2.12 Blast furnace gases Gaseous fuels produced in the production of iron in the blast furnace.

6.2.13 Water gases Gases produced by reacting coke with steam.

6.2 Types de Gaz

6.2.1 Gaz combustibles Gaz ou mélanges de gaz qui sont combustibles en mélange avec l'air ou avec l'oxygène et qui sont utilisés notamment pour la production de chaleur.

6.2.2 Familles de gaz Constituées de gaz combustibles dont les caractéristiques de combustion sont très proches les unes des autres, de telle sorte qu'elles permettent leur interchangeabilité. La première famille de gaz comprend le gaz de ville et le gaz de réseau de transport (gaz combustibles riches en hydrogène). La deuxième famille comprend les gaz naturels, les gaz associés au pétrole ainsi que les gaz interchangeables avec eux. La troisième famille de gaz comprend les gaz de pétrole liquéfiés (propane, butane). Observation : les familles de gaz peuvent être définies par une plage de l'indice Wobbe (ou du pouvoir calorifique supérieur ou inférieur.)

6.2.3 Gaz naturels Gaz combustibles riches en méthane qui proviennent de gisement naturels.

6.2.4 Gaz associés au pétrole Gaz combustibles riches en méthane provenant de gisements naturels dont la plus importante fraction peut être constituée par des hydrocarbures supérieurs (par exemple, gaz en poche, gaz dissous).

6.2.5 Gaz de pétrole liquéfiés Hydrocarbures en C3 et C4 et leurs mélanges.

6.2.6 Gaz de raffinerie Gaz combustibles produits lors du raffinage du pétrole.

6.2.7 Gaz de cokerie Gaz combustibles produits au cours de la fabrication du coke.

6.2.8 Gaz de gazéification sous pression Gaz combustibles produits à partir de combustibles solides, liquides ou gazeux, au moyen de gazéification (par exemple, un mélange oygène/vapeur) et sous haute pression.

6.2.9 Gaz de craquage Gaz combustibles produits par transformation thermique ou thermocatalytique d'hydrocarbures liquides ou gazeux.

6.2.10 Gaz de ville Gaz combustibles de la 1ère famille de gaz avec un indice Wobbe comprise entre 24,8-28,8 MJ/m³.

6.2.11 Gaz de gazogène Gaz combustibles produits principalement à partir de combustibles solides par une gazéification réalisée au moyen d'air ou d'air saturé en vapeur d'eau.

6.2.12 Gaz de hauts-fourneaux Gaz combustibles produits lors de la fabrication de la fonte au haut-fourneau.

6.2.13 Gaz à l'eau, gaz bleu Gaz combustibles obtenus par gazéification de coke au moyen de vapeur d'eau.

6.2 Gasarten

6.2.1 Brenngase Brenngase sind Gase oder Gasgemische, die in Mischung mit Luft oder Sauerstoff brennbar sind und vorwiegend für die Wärmeerzeugung eingesetzt werden.

6.2.2 Gasfamilien Gasfamilien sind Brenngase mit weitgehend übereinstimmenden Brenneigenschaften, die ihre Austauschbarkeit ermöglichen. Die 1. Gasfamilie umfasst Stadt- und Ferngase (wasserstoffreiche Brenngase), die 2. Gasfamilie Erdgase, Erdölbegleitgase sowie deren Austauschgase und die 3. Gasfamilie die Flüssiggase (Propan, Butan).

Die Gasfamilien können durch einen Bereich des Wobbeindexes (oder des Brenn- oder Heizwerkes) definiert werden.

6.2.3 Erdgase Erdgase sind aus natürlichen Lagerstätten stammende methanreiche Brenngase.

6.2.4 Erdölbegleitgase Erdölbegleitgase sind aus natürlichen Vorkommen stammende methanreiche Brenngase, die grössere Anteile an höheren Kohlenwasserstoffen enthalten können. (z.B. Kappengase, gelöste Gase usw.)

6.2.5 Flüssiggase Flüssiggase sind C3- und C4-Kohlenwasserstoffe sowie deren Gemische.

6.2.6 Raffineriegase Raffineriegase sind Brenngase, die bei der Verarbeitung von Erdöl anfallen.

6.2.7 Kokereigase Kokereigase sind Brenngase, die bei der Herstellung von Koks entstehen.

6.2.8 Druckvergasungsgase Druckvergasungsgase sind Brenngase, die aus festen, flüssigen oder gasförmigen Brennstoffen mit einem Vergasungsmittel (z.B. Sauerstoff-Dampf-Gemisch) unter Hochdruck erzeugt werden.

6.2.9 Spaltgase Spaltgase sind Brenngase, die aus flüssigen oder gasförmigen Kohlenwasserstoffen durch thermische oder thermisch-katalytische Umsetzung hergestellt werden.

6.2.10 Stadtgase Stadtgase sind Brenngase der 1. Gasfamilie mit einer Wobbezahl zwischen 24,8-28,8 MJ/m^3.

6.2.11 Generatorgase Generatorgase sind Brenngase, die vorwiegend aus festen Brennstoffen durch Vergasung mittels Luft oder wasserdampfgesättigter Luft erzeugt werden.

6.2.12 Gichtgase Gichtgase sind Brenngase, die bei der Gewinnung von Roheisen im Hochofen anfallen.

6.2.13 Wassergase Wassergase sind Brenngase, die durch Vergasung von Koks mit Wasserdampf entstehen.

6.2 Clases de Gas

6.2.1 Gases combustibles Gas o mezclas de gas que mezcladòs con el aire o con el oxígeno, son combustibles y que se utilizan principalmente para la utilización de calor.

6.2.2 Familias de gases Están constituidas por gases combustibles cuyas características de combustión están my próximas entre si lo que permite que sean intercambiables. La primera familia de gases comprende el gas ciudad y los gases de red de transporte (gases combustibles ricos en hidrógeno). La segunda familia comprende los gases nautrales, los gases asociados al petróleo así como los gases intercambiables con ellos. La tercera familia de gases comprende los gases licuados de petróleo (propano, butano).

Las familias de gases pueden definirse por el indice Wobbe o por su poder calorifico superior e inferior.

6.2.3 Gases naturales Gases naturales ricos en metano y que proceden de yacimientos naturales.

6.2.4 Gases asociados al petróleo Gases combustibles ricos en metano que proceden de yacimientos naturales, cuya fracción más importante puede - estar constituida por hidrocarburos superiores.

6.2.5 Gases licuados de petróleo (GLP) Hidrocarburos en C-3 y C-4 y sus mezclas (ver sección 5).

6.2.6 Gases de refinería Gases producidos durante el refino y tratamiento del petróleo.

6.2.7 Gases de coquerías Gases combustibles producidos durante la fabricación del coque.

6.2.8 Gases de gasificación a alta presión Gases combustibles producidos a partir de combustibles sólidos, líquidos o gaseosos, mediante un agente gasificante (p. ej. mezcla de oxígeno y vapor de agua) a alta presión.

6.2.9 Gases de craqueo Gases combustibles producidos a partir de hidrocarburos líquidos o gaseosos por transformación térmica o termocatalítica.

6.2.10 Gas de ciudad Gases combustibles comprendidos en la primera familia de gases, con un índice de Wobbe comprendido entre 24,8-28,8 MJ/m^3.

6.2.11 Gases de gasógeno Gases combustibles producidos principalmente por gasificación contínua de combustibles sólidos mediante aire o una mezcla de aire saturado en vapor de agua.

6.2.12 Gases de horno alto Gases combustibles producidos durante la obtención de hierro en el horno alto.

6.2.13 Gas de agua, gas azul Gases combustibles obtenidos por gasificación del coque por medio de vapor de agua.

6

6.2.14 Synthetic natural gas (SNG) Gaseous fuel manufactured from coal or hydrocarbons, or from other carbonaceous material and interchangeable with natural gas.

6.3 Natural Gas Production

6.3.1 Natural gas deposit/reservoir/pool/producing formation (USA), **pay horizon** (USA) A natural accumulation of gaseous hydrocarbons in underground porous rocks or caverns.

6.3.2 Gas-bearing stratum; gas stratum A gas-bearing porous stratum or cavern-containing stratum within a natural gas or petroleum deposit.

6.3.3 Natural gas field One or more reservoirs grouped in or related to the same individual geological structured feature. In some countries the term relates to the surface area above a deposit on which is located the equipment for extracting, treating, transporting, etc., the natural gas.

6.3.4 Natural gas production; natural gas extraction The application of industrial technology to bring the natural gas to the surface from the deposit.

6.4 Gas Manufacture

6.4.1 Carbonisation The heating under controlled conditions and in the absence of air of solid fuels to produce gaseous, liquid and solid products.

6.4.2 Coking The heating under controlled conditions and in the absence of air of solid fuels at temperatures in excess of 900°C to produce coke.

6.4.3 Gasification The conversion of solid or liquid fuels to gaseous fuels by reaction with a gasification medium such as steam, air or oxygen; it can also be the conversion of liquid fuels to gaseous fuels by thermal or catalytic processes. It may be conducted at atmospheric, medium or high pressure. See note to 6.4.4.

6.4.4 Gasification under pressure; pressure gasification The conversion of solid or liquid fuels to gaseous fuels by reaction with a gasification medium (e.g. oxygen/steam mixture) under high pressure. It can also be the conversion of liquid fuels to gaseous fuels by thermal or catalytic processes at high pressure.

Note to 6.4.3 and 6.4.4 above In processes in which a gas is the feedstock the process is generally referred to by a more specific term, e.g. reforming, methanation.

6.4.5 Cracking The production of gaseous fuels by the thermal or thermal-catalytic conversion of liquid or gaseous fuels.

6.4.6 Conversion; shift reaction A process for reducing the CO-content of a gaseous fuel by catalytically converting it with steam to CO_2 and H_2.

6.5 Gas Processing

6.5.1 Purification of fuel gas/gaseous fuel/combustible gas The removal of impurities from the gas.

6.5.2 Water removal; demisting A process for removing condensed water from natural gas.

6.5.3 Sulphur removal; desulphurization; desulphurizing process A process for removing sulphur compounds contained in gaseous fuels.

6.2.14 Gaz naturel synthétique Gaz combustible obtenu à partir de charbon ou d'hydrocarbure interchangeable avec le gaz naturel.

6.3 Exploitation du Gaz Naturel

6.3.1 Gisement de gaz naturel Accumulation naturelle d'hydrocarbures en phase gazeuse se trouvant dans des couches poreuses ou caverneuses de l'écorce terrestre.

6.3.2 Couche gazéifère Couche poreuse ou caverneuse contenant du gaz naturel, comprise à l'intérieur d'un gisement de gaz naturel ou de pétrole.

6.3.3 Champ de gaz naturel Zone délimitée de la surface terrestre se trouvant au-dessus d'un gisement de gaz naturel et sur laquelle sont installées et en service des installations d'extraction, de transport et, éventuellement, de traitement.

6.3.4 Extraction du gaz naturel Consiste à amener au jour le gaz naturel provenant d'un gisement au moyen de dispositifs industriels.

6.4 Production du Gaz Manufacture

6.4.1 Carbonisation Procédé de traitement thermique d'un combustible solide en l'absence d'air.

6.4.2 Cokefaction Procédé de traitement thermique d'un combustible solide en l'absence d'air à des températures supérieures à 900°C.

6.4.3 Gazéification Procédé de fabrication de gaz combustibles par réaction de combustibles solides ou liquides avec un agent de gazéification, par exemple de l'air ou de l'oxygène, à la pression atmosphérique.

6.4.4 Gazéification sous pression Procédé de production de gaz combustibles à partir de combustibles solides, liquides ou gazeux par une réaction faisant intervenir un agent de gazéification, par exemple un mélange oxygène/vapeur, et sous haute pression.

6.4.5 Craquage Procédé de fabrication de gaz combustibles à partir de combustibles liquides ou gazeux au moyen d'une transformation thermique ou thermocatalytique.

6.4.6 Conversion La conversion est un procédé permettant de diminuer la teneur en CO d'un gaz combustible par une transformation catalytique en présence de vapeur d'eau.

6.5 Traitement du Gaz

6.5.1 Epuration (d'un gaz combustible) Opération qui consiste à éliminer des fractions gênantes des gaz combustibles.

6.5.2 Opération de séparation de l'eau Opération qui consiste à éliminer l'eau condensée contenue dans un gaz naturel.

6.5.3 Désulfuration, épuration Opération qui consiste à éliminer les composés du soufre contenus dans des gaz combustibles.

6.2.14 Synthetisches Erdgas (SNG) Synthetisches Erdgas wird aus Kohle oder Erdöl erzeugt, mit den gleichen Eigenschaften wie Erdgas, mit dem es austauschbar ist.

6.3 Erdgasbereitstellung

6.3.1 Erdgaslagerstätte Erdgaslagerstätte ist eine natürliche Akkumulation von gasförmigen Kohlenwasserstoffen in porösen oder kavernösen Schichten der Erdrinde.

6.3.2 Gasführende Schicht Gasführende Schicht ist die innerhalb einer Erdgas- oder Erdöllagerstätte gelegene erdgashaltige poröse oder kavernöse Schicht.

6.3.3 Erdgasfeld Erdgasfeld ist ein auf einer Erdgaslagerstätte liegendes abgegrenztes Gebiet der Erdoberfläche, auf dem Förder-, Transport- und ggf. Aufbereitungsanlagen installiert und betrieben werden.

6.3.4 Erdgasförderung Erdgasförderung ist die Zutagebringung des Erdgases aus der Lagerstätte mittels technischer Einrichtungen.

6.4 Gaserzeugung

6.4.1 Entgasung Entgasung ist ein Verfahren der thermischen Behandlung eines festen Brennstoffes unter Luftabschluss.

6.4.2 Verkokung Verkokung ist ein Verfahren der thermischen Behandlung eines festen Brennstoffes unter Luftabschluss bei Temperaturen oberhalb 900°C.

6.4.3 Vergasung Vergasung ist ein Verfahren zur Herstellung von Brenngasen durch Reaktion fester oder flüssiger Brennstoffe mit einem Vergasungsmittel, z.B. Luft oder Sauerstoff unter atmosphärischem Druck.

6.4.4 Druckvergasung Druckvergasung ist ein Verfahren zur Gewinnung von Brenngasen aus festen, flüssigen oder gasförmigen Brennstoffen durch Reaktion mit einem Vergasungsmittel z.B. Sauerstoff-Dampf-Gemisch unter Hochdruck.

6.4.5 Spaltvergasung Spaltvergasung ist ein Verfahren zur Herstellung von Brenngasen aus flüssigen oder gasförmigen Brennstoffen mittels thermischer oder thermisch-katalytischer Umsetzung.

6.4.6 Konvertierung Konvertierung ist ein Verfahren zur Verringerung des CO-Gehaltes im Brenngas durch katalytische Umsetzung mit Wasserdampf.

6.5 Gasaufbereitung

6.5.1 Brenngasreinigung Brenngasreinigung ist ein Verfahren zum Entfernen störender Bestandteile aus Brenngasen.

6.5.2 Entwässerung Entwässerung ist ein Verfahren zum Entfernen des im geförderten Erdgas enthaltenen tropfbaren Wassers.

6.5.3 Entschwefelung Entschwefelung ist ein Verfahren zum Entfernen der in Brenngasen enthaltenen Schwefelverbindungen.

6.2.14 Gases naturales sintéticos Gases combustibles obtenidos a partir del carbón o de hidrocarburos, intercambiables con el gas natural.

6.3 Producción de Gas Natural

6.3.1 Yacimiento de gas natural Acumulación natural de hidrocarburos gaseosos en formaciones de rocas porosas subterráneas o en cavernas de la corteza terrestre.

6.3.2 Capa gasífera Estrato poroso o cavernoso, que contiene gas natural que se encuentra en el interior de un yacimiento de gas natural o de petróleo.

6.3.3 Campo de gas natural Zona delimitada de la superficie terrestre que se encuentra sobre un yacimiento de gas natural y sobre la que se encuentran instaladas y en servicio las instalaciones de extracción, transporte y, eventualmente, de tratamiento.

6.3.4 Extracción del gas natural Consiste en hacer llegar a la superficie el gas natural, procedente de un yacimiento, por medio de dispositivos industriales.

6.4 Fabricación del Gas

6.4.1 Carbonización Procedimiento de tratamiento térmico de un combustible sólido en ausencia de aire.

6.4.2 Coquización Procedimiento de tratamiento térmico de un combustible sólido en ausencia de aire, a temperaturas superiores a 900° C.

6.4.3 Gasificación Procedimiento de fabricación de gases combustibles por reacción de combustibles sólidos o líquidos con un agente gasificante, por ejemplo el aire o el oxígeno a la presión atmosférica.

6.4.4 Gasificación a presión Procedimiento de producción de gases combustibles a partir de combustibles sólidos líquidos o gaseosos por medio de una reacción que hace intervenir un agente gasificante, por ejemplo una mezcla oxígeno/vapor, a alta presión.

6.4.5 Craqueo Procedimiento de fabricación de gases combustibles a partir de combustibles líquidos o gaseosos por medio de una transformación térmica o termocatalítica.

6.4.6 Conversión Procedimiento que permite disminuir el contenido en CO de un gas combustible por medio de una transformación catalítica en presencia de vapor de agua.

6.5 Tratamiento del Gas

6.5.1 Depuración (de un gas combustible) Operación que consiste en eliminar las impurezas de los gases combustibles.

6.5.2 Separación del agua Operación que consiste en eliminar el agua condensada contenida en un gas natural.

6.5.3 Desulfuración Operación que consiste en eliminar los compuestos de azufre contenidos en los gases combustibles.

6

6.5.4 Gasoline stripping The removal of gasoline fractions contained in liquid and vapour form in natural gas during production.

6.5.5 Dehydration The removal of water vapour from gaseous fuels.

6.5.6 Conditioning A process of adjusting the characteristics of a gaseous fuel as required, by the admixture of other gases or liquids. In the USA the term embraces both the removal of objectionable constituents and the addition of desirable constituents.

6.5.7 Enrichment Raising the calorific value of a gas by mixing with it a gas of relatively high heating value.

6.5.8 Liquefaction The conversion of natural gas to the liquid phase.

6.5.9 Odorization; odorizing The addition of a liquid chemical substance in the vapour phase to a gas so that it becomes identifiable by its disagreeable smell; this can be a legal requirement.

6.6 Gas Transmission and Distribution

6.6.1 Gas transmission line; gas pipeline A pipeline for the transmission of gaseous fuel at high pressure and over long distances; the term normally includes the ancillary equipment.

6.6.2 Gas pipeline crossing one frontier; international pipeline; interstate pipeline
Note. There is no specific term in English denoting a pipeline that crosses one frontier only.

6.6.3 Gas pipeline crossing two or more frontiers; international pipeline; interstate pipeline
Note. There is no specific term in English denoting a pipeline that crosses two or more frontiers.

6.6.4 Transmission and distribution system/network/grid The whole of the pipelines and mains, including associated components, such as pipe fittings, valves, connections, house branch mains, pig traps, etc.

6.6.5 Compressor A machine in which the pressure or velocity of a gas is increased for the purpose of transmitting or storing it.

6.6.6 Compressor plant/installation/station Plant for compressing gas, comprising compressors, compressor motive power, metering, regulation and control equipment, associated piping and ancillary equipment, safety equipment, civil engineering works.

6.6.7 Distribution system/network/grid The system of gas mains that provides for the local distribution of gaseous fuel.

6.6.8 Gas pressure regulator station; gas governor station A plant that automatically reduces a higher gas pressure to a constant lower value.

6.6.9 Gas pressure regulator; gas governor A device that automatically reduces a higher gas pressure to a constant lower value.

6.6.10 Gas meter An instrument with an indicating mechanism that directly measures volumes of gas.

6.6.11 Tanker A merchant ship designed to transport liquid cargoes, e.g. liquefied natural gas; in context the more specific terms methane tanker, propane tanker, butane tanker would be used.

6.5.4 Dégazolinage, désessenciement Opération qui consiste à éliminer la fraction essence ou gazoline, en phase liquide et vapeur, contenue dans un gaz naturel.

6.5.5 Séchage, déshydratation Opération qui consiste à éliminer la vapeur d'eau contenue dans des gaz combustibles.

6.5.6 Conditionnement Opération qui consiste à conférer à un gaz combustible des caractéristiques déterminées, par mélange avec d'autres gaz ou de liquides.

6.5.7 Enrichissement Procédé permettant d'éliminer les fractions inertes d'un gaz naturel ou de lui ajouter des hydrocarbures.

6.5.8 Liquéfaction Opération qui consiste à transformer le gaz naturel en une phase liquide.

6.5.9 Odorisation Opération qui consiste à mélanger à des gaz combustibles des composés liquides en phase vapeur d'odeur désagréable, pour permettre leur détection.

6.6 Transport du Gaz

6.6.1 Conduite de transport (de gaz) gazoduc Conduite assurant le transport d'un gaz combustible sous haute pression et à longue distance.

6.6.2 Conduite d'exportation (de gaz) Conduite de transport de gaz qui traverse une seule frontière d'état.

6.6.3 Conduite de transit (de gaz) Conduite de transport de gaz combustible à haute pression qui traverse au moins deux frontières.

6.6.4 Installation de transport ou de distribution, réseau Ensemble des conduites et des accessoires correspondants, comme : pièces de raccords, robinetterie, prises de branchement, branchements proprement dits, sas pour piston-râcleur, etc. . .

6.6.5 Compresseur Machine dans laquelle l'énergie d'un gaz, sous forme de pression ou de vitesse, est augmentée en vue de son transport ou de son stockage.

6.6.6 Station de compression Installation destinée à la compression d'un gaz, se composant de compresseurs, de dispositifs accessoires de mesurage, de contrôle et de régulation, d'installations de distribution d'énergie, de tuyauteries, d'installations annexes, de dispositifs de sécurité et d'ouvrages de génie civil.

6.6.7 Réseau de distribution (de gaz) Réseau de canalisations ayant pour objet la distribution locale d'un gaz combustible.

6.6.8 Poste de détente (du gaz) Installation fonctionnant automatiquement, ayant pour but d'abaisser le pression du gaz à une valeur plus faible et constante.

6.6.9 Détenteur-régulateur de gaz Appareil fonctionnant automatiquement qui a pour effet d'abaisser la pression du gaz à une valeur plus faible et constante.

6.6.10 Compteur de gaz Appareil comprenant un dispositif indicateur assurant le mesurage direct des quantités de gaz.

6.6.11 Bateau-citerne Navire de charge destiné au transport de cargaisons liquides, par exemple du gaz naturel liquéfié. Suivant la charge, on peut distinguer par exemple : le butanier, le propanier, le méthanier.

6.5.4 Entbenzinierung Entbenzinierung ist ein Verfahren zum Entfernen der im geförderten Erdgas enthaltenen flüssigen und dampfförmigen Benzinanteile.

6.5.5 Trocknung Trocknung ist ein Verfahren zum Entfernen des in Brenngasen enthaltenen Wasserdampfes.

6.5.6 Konditionierung Konditionierung ist ein Verfahren zum Einstellen vorgegebener Qualitätsmerkmale eines Brenngases durch Zumischen von Fremdgasen oder Flüssigkeiten.

6.5.7 Anreicherung Anreicherung ist ein Verfahren zum Entfernen inerter Bestandteile aus einem Erdgas oder der Zumischung von Kohlenwasserstoffen.

6.5.8 Verflüssigung Verflüssigung ist ein Verfahren der Ueberführung des gasförmigen Erdgases in die flüssige Phase.

6.5.9 Odorierung Odorierung ist Beimischen unangenehm riechender Flüssigkeiten in Dampfform zu Brenngasen zum Zwecke der Indikation.

6.6 Gastransport

6.6.1 Gasfernleitung Gasfernleitung ist eine Rohrleitung zum Transport von Brenngasen unter Hochdruck über weite Entfernung.

6.6.2 Grenzüberschreitende Gasleitung Grenzüberschreitende Gasleitung ist eine Gasrohrleitung, die nur eine Staatsgrenze überschreitet.

6.6.3 Transitgasleitung Transitgasleitung ist eine mindestens zweifach grenzüberschreitende Rohrleitung zum Transport von Brenngasen unter Hochdruck.

6.6.4 Rohrleitungsanlage Rohrleitungsanlage ist die Gesamtheit der Rohrleitungen einschliesslich zugehörigen Zubehörs, wie Formstücke, Armaturen, Anschlussstücke, Hausanschlussleitungen, Molchschleusen usw.

6.6.5 Verdichter Verdichter ist eine Arbeitsmaschine, in der die Druck- oder Geschwindigkeitsenergie eines Gases erhöht wird, um es zu fördern oder zu speichern.

6.6.6 Verdichteranlage Verdichteranlage ist eine Anlage zum Verdichten von Gasen, bestehend aus den Verdichtern, den zugehörigen MSR-Einrichtungen, Energieversorgungsanlagen, Rohrleitungen, Nebenanlagen, Sicherheitsvorrichtungen und baulichen Einrichtungen.

6.6.7 Gasverteilungsnetz Gasverteilungsnetz ist ein Rohrleitungssystem zur örtlichen Verteilung von Brenngasen.

6.6.8 Gasdruckregelanlage Gasdruckregelanlage ist eine selbsttätig wirkende Anlage zum Herabsetzen eines höheren Gasdruckes auf einen konstanten niederen Wert.

6.6.9 Gasdruckregelgerät Gasdruckregelgerät ist ein selbsttätig wirkendes Gerät zum Herabsetzen eines höheren Gasdruckes auf einen konstanten niederen Wert.

6.6.10 Gaszähler Gaszähler ist ein Gerät mit Zählwerk zur direkten Messung von Gasmengen.

6.6.11 Tanker Tanker ist ein Frachtschiff zum Transport flüssiger Ladung, z.B. Flüssigerdgas. Je nach Ladung werden Methan-, Propan- und Butantanker unterschieden.

6.5.4 Desgasolinado Operación que consiste en eliminar la fracción de gasolina en fase líquida y de vapor, contenida en un gas natural.

6.5.5 Secado, deshidratación Operación que consiste en eliminar el vapor de agua, contenido en los gases combustibles.

6.5.6 Acondicionamiento Operación que consiste en proporcionar a un gas combustible determinadas características por mezcla con otros gases o con líquidos.

6.5.7 Enriquecimiento Procedimiento para eliminar las fracciones inertes de un gas o para agregarle hidro-carburos.

6.5.8 Licuación Operación que consiste en transformar el gas natural en una fase líquida.

6.5.9 Odorización Operación que consiste en mezclar con los gases combustibles compuestos líquidos de olor desagradable, en fase de vapor, para permitir su detección.

6.6 Transporte del Gas

6.6.1 Conducción para transporte (de gas); gasoducto Conducción que asegura el transporte de un gas combustible, a alta presión a larga distancia.

6.6.2 Gasoducto internacional Conducción para el transporte de gas que atraviesa una sola frontera entre dos estados.

6.6.3 Gasoducto de tránsito Conducción de transporte de gas combustible a alta presión que atraviesa, por lo menos, dos fronteras.

6.6.4 Instalación de transporte o de distribución, red Conjunto de conducciones y de los accesorios correspondientes tales como elementos de unión, válvulas, tomas de acometidas, acometidas propiamente dichas, trampas para los pistones, rascadores etc.

6.6.5 Compresor Máquina en la que la energía de un gas, en forma de presión o de velocidad es aumentada con vistas a su transporte o almacenamiento.

6.6.6 Instalación de compresión Instalación destinada a la compresión de un gas compuesta de compresores, dispositivos, accesorios de medida de control y de regulación, instalaciones de distribución de energía, tuberías, instalaciones anexas, dispositivos de seguridad y obras de ingeniería civil.

6.6.7 Red de distribución (de gas) Red de canalizaciones que tiene por objeto la distribución local de un gas combustible.

6.6.8 Estación para regulación de la presión del gas Instalación de funcionamiento automático, que tiene por objeto reducir la presión del gas a un valor más bajo, constante.

6.6.9 Regulador de presión del gas Aparato de funcionamiento automático cuyo efecto es el de reducir la presión de gas a un valor más bajo, constante.

6.6.10 Contador de gas Aparato que cuenta con un dispositivo indicador que asegura la medición directa de las cantidades de gas.

6.6.11 Buque cisterna Nave de carga destinada al transporte de cargamentos líquidos por ejemplo, gas licuado natural. Según la clase de cargas se denominan p. ej. butanero, propanero, metanero.

6

6.6.12 Rail tanker; rail tank car A railway freight car for the transport of liquids, e.g. liquefied petroleum gases.

6.6.13 Road tanker; tank truck (USA) A road vehicle for the transport of liquids; e.g. liquefied petroleum gases.

6.6.14 Transportable gas holder; transportable gas container; gas cylinder A container in bottle, spherical or cylindrical form for the transport and distribution of liquefied gases.

6.7 Gas Storage

6.7.1 Underground gas storage; underground gas storage system Storage in porous geological formations, natural or artificially created cavities, suitable for the storage of gaseous fuels.

6.7.2 Storage in porous rock Storage in a porous rock formation suitable for the storage of gas; examples are storage in aquifers, depleted gas wells or reservoirs.

6.7.3 Storage in underground cavities Underground storage in natural or artificial integral cavities; examples are storage in saline cavities, natural caverns, disused mine workings, frozen earth.

6.7.4 Storage in caverns Storage in cavities artificially created by washing out water soluble layers of rock, e.g. rock salt.

6.7.5 Storage in fissures Storage in cavities of a kind suitable for gas storage, in which the reservoir rock is very fissured due to tectonic stresses.

6.7.6 Current gas; active gas The quantity of gas available within the storage range of an underground gas storage, that serves to balance out the differences between gas available in the system and demand.

6.7.7 Cushion gas The quantity of gas associated with gas storage that can never be completely recovered.

6.7.8 Gas holder A vessel in which gas is stored at or near the surface in gaseous or liquid phase.

6.7.9 Low-pressure gas holder A general term for bell-type, piston or waterless gas holders.

6.7.10 Bell-type gas holder A hollow cylinder closed at its upper end and sealed at its lower end by a liquid, generally water, contained in a tank; the gas is stored at low pressure within the cylinder above the level of the water; the cylinder, being free to rise or fall, is able to accommodate a varying volume of gas.

6.7.11 Piston type gas holder; waterless gas holder; dry gas holder A tall vessel, polygonal or circular in plan, inside which a disc or piston, having a gas-tight sliding joint or flexible diaphragm connection with the vessel, is free to move vertically; the gas is stored in the space beneath the disc or piston, at low pressure.

6.7.12 High-pressure gas holder; pressure type gas holder; pressure holder A fixed or movable, closed vessel of constant volume in which gas is stored at a pressure of several atmospheres.

6.6.12 Wagon-citerne Wagon de transport de marchandises par voie ferrée, destiné au transport de liquides, par exemple des gaz de pétrole liquéfiés.

6.6.13 Camion-citerne, citerne routière Véhicule routier destiné au transport de chargements liquides, par exemple des gaz de pétrole liquéfiés.

6.6.14 Récipient de transport Récipient en forme de bouteille, de sphère ou de cylindre, destiné au transport et à la distribution de gaz liquéfiés.

6.7 Stockage du Gaz

6.7.1 Stockage souterrain (de gaz) Couches géologiques poreuses ou encore cavités naturelles ou créées artificiellement qui conviennent pour le stockage des gaz combustibles.

6.7.2 Stockage en couche poreuse Couche de roche poreuse qui convient pour le stockage du gaz. Il peut s'agir de "nappe aquifère" ou de "gisement épuisé".

6.7.3 Stockage en excavation souterraine Stockage souterrain constitué par une cavité naturelle ou artificielle et présentant une cohésion suffisante. Il peut s'agir de cavités salines, de cavernes naturelles, de galeries minières ou de "terre gelée".

6.7.4 Stockage en cavité saline Stockage en cavité créée artificiellement, par lessivage d'une couche rocheuse soluble dans l'eau (par exemple : sel gemme).

6.7.5 Stockage en couche fracturée Stockage dans des excavation souterraines convenables, dont la roche est fortement fracturée par suite de contraintes tectoniques.

6.7.6 Gaz utile, respiration Quantité de gaz disponible à l'intérieur de l'espace actif d'un stockage souterrain qui doit servir à compenser la différence entre les disponibilités du réseau et la consommation.

6.7.7 Gaz coussin Quantité de gaz emmagasinée dans un stockage souterrain qui ne peut pas être complètement récupérée après son injection.

6.7.8 Réservoirs (de gaz) Dispositifs en surface ou proches du sol, destinés au stockage de gaz combustibles en phase gazeuse ou liquide.

6.7.9 Réservoirs (de gaz) à basse pression Terme général qui désigne les gazomètres hydrauliques et les gazomètres secs.

6.7.10 Gazomètre hydraulique, à cloche, à cuve Réservoir utilisé pour le stockage de gaz à basse pression et dont la capacité de stockage est formée par une cloche généralement à plusieurs levées et plongeant dans une cuve.

6.7.11 Gazomètre sec Corps creux fixe dont la surface de base est de forme circulaire ou polygonale, fermé à la partie supérieure par un piston mobile et destiné au stockage du gaz à basse pression.

6.7.12 Réservoirs (de gaz) sous pression Réservoirs fixes ou mobiles qui peuvent être remplis d'un gaz combustible sous pression.

6.6.12 Kesselwagen Kesselwagen ist ein Eisenbahngüterwagen zum Transport von Flüssigkeiten, z.B. von Flüssiggasen.

6.6.13 Tankwagen Tankwagen ist ein Strassenfahrzeug zum Transport flüssiger Ladung, z.B. Flüssiggas.

6.6.14 Transportbehälter Transportbehälter ist ein Behälter in Flaschen-, Kugel-, oder Zylinderform zum Transport und Verteilung von verflüssigten Gase.

6.7 Gasspeicherung

6.7.1 Gasspeicher Gasspeicher sind für die Bevorratung von Brenngasen geeignete poröse geologische Gesteinsschichten sowie natürliche oder künstlich geschaffene Hohlräume.

6.7.2 Porenspeicher Porenspeicher ist eine für die Gasspeicherung geeignete poröse Gesteinsschicht. Es kann sich um Aquiferspeicher oder erschöpfte Lagerstätten handeln.

6.7.3 Hohlraumspeicher Hohlraumspeicher ist e n unterirdischer Speicher, der aus einem natürlichen oder künstlich geschaffenen, zusammenhängenden Hohlraum besteht, z.B. Salzkaverne, natürliche Kaverne, Grubenbau oder Gefrierspeicher.

6.7.4 Kavernenspeicher Kavernenspeicher ist ein künstlich geschaffener Hohlraumspeicher, der durch Aussolen wasserlöslicher Gesteinsschichten (z.B. Steinsalz) entsteht.

6.7.5 Kluftspeicher Kluftspeicher ist ein für die Gasspeicherung geeigneter Hohlraumspeicher, dessen Speichergestein infolge tektonischer Beanspruchungen stark zerklüftet ist.

6.7.6 Aktivgas Aktivgas ist eine innerhalb des Speichermengenspiels eines unterirdischen Gasspeichers verfügbare Gasmenge, die dem Ausgleich der Differenz zwischen Bereitstellung und Verbrauch dient.

6.7.7 Kissengas Kissengas ist eine im Speicher gebundene Gasmenge, die nach dessen Stillegung nicht vollständig zurückgewonnen werden kann.

6.7.8 Gasbehälter Gasbehälter sind oberirdische oder oberflächennahe Anlagen zur Speicherung von Brenngasen im gasförmigem oder verflüssigtem Zustand.

6.7.9 Niederdruckgasbehälter Niederdruckgasbehälter ist ein Sammelbegriff für Glocken- und Scheibengasbehälter.

6.7.10 Glockengasbehälter Glockengasbehälter ist ein Behälter, der zur Speicherung von Gas unter Niederdruck dient und dessen Speicherraum von einer in ein Waserbecken eintauchenden, meist unterteilten Glocke gebildet wird.

6.7.11 Scheibengasbehälter Scheibengasbehälter ist ein stehender Hohlkörper mit kreisförmiger oder polygonaler Grundfläche zur Speicherung von Gas unter Niederdruck, der nach oben von einer beweglichen Scheibe abgedichtet wird.

6.7.12 Druckgasbehälter Druckgasbehälter sind ortsfeste oder ortsbewegliche Behälter, die mit Brenngas unter Druck gefüllt werden können.

6.6.12 Vagón cisterna Vagón de transporte por vía férrea, destinado al transporte de líquidos, por ejemplo de gases licuados de petróleo.

6.6.13 Camión-cisterna Vehículo destinado al transporte, por carretera, de cargamentos líquidos, por ejemplo gas licuado de petróleo.

6.6.14 Recipientes para transporte, bombonas Recipientes en forma de botella, esfera o cilindro, destinado al transporte y distribución de gases licuados.

6.7 Almacenamiento del Gas

6.7.1 Almacenamiento subterráneo (de gas) Capas geológicas porosas o también cavidades naturales o artificiales, adecuadas para el almacenaje de gases combustibles.

6.7.2 Almacenamiento en capas porosas Capa de roca porosa adecuada para el almacenamiento de gas. Puede tratarse de una capa acuífera o de un yacimiento agotado.

6.7.3 Almacenamiento en cavidades Almacenamiento subterráneo que se realiza en espacios huecos naturales o artificiales, que presentan suficiente cohesión. Puede tratarse de cavidades salinas, cavernas naturales, galerías mineras o tierras congeladas.

6.7.4 Almacenamiento en cavidad salina Almacenamiento en una cavidad creada artificialmente por lixiviación (disolución) de una capa rocosa soluble en agua (por ejemplo: sal gema).

6.7.5 Almacenamiento en capas fracturadas Almacenamiento subterráneo que se realiza en una roca almacén muy fisurada a causa de los esfuerzos tectónicos.

6.7.6 Gas activo, gas de respiración Cantidad de gas disponible en el interior del espacio útil de un almacenamiento subterráneo y que debe servir para compensar la diferencia entre las disponibilidades de la red y el consumo.

6.7.7 Colchón de gas Es la cantidad de gas depositado en un almacenamiento subterráneo que no puede recuperarse completamente después de haber sído inyectado.

6.7.8 Depósito (de gas) Dispositivo elevado o próximo al terreno destinado al almacenaje de gas combustible en fase gaseosa o líquida.

6.7.9 Gasómetro de gas de baja presion Expresión general que designa los gasómetros de campana y los gasómetros secos.

6.7.10 Gasómetro hidráulico, de campana de cuba Depósito utilizado para el almacenamiento de gas a baja presión y cuya capacidad de almacenamiento está formada por una campana generalmente con varias elevaciones y sumergidas en una cuba.

6.7.11 Gasómetro seco Cuerpo cóncavo fijo, cuya superficie de base tiene forma circular o poligonal, cerrado en su parte superior por un pistón móviles que pueden estar llenos de un gas combaja pressión.

6.7.12 Depósito de gas a alta presión Depósitos fijos o móviles que pueden estar llenos de un gas combustible a presión.

6

6.8 Gas Utilisation

6.8.1 **Service connection** The branch lines between the supply mains and the main service valve or meter cock, the insulating joint, the main service valve or meter cock itself, the service valve or cock located outside the building, if any, and the gas governor of the building.

6.8.2 **Service pipe; house branch line; house lateral (USA); house dead end line (USA); domestic mains** (Aust.) The pipe connecting the supply main to the service valve or meter cock.

6.8.3 **Consumer's plant; customer's plant** A plant that takes gas from the public supply downstream of the main service valve or meter cock.

6.8.4 **Installed capacity; connected load** The sum of the rated heat inputs of the gas-consuming appliances connected to the supplying system or any part of the system under consideration, e.g. the appliances of a consumer.

Note. A consumer's *contractual demand* would be at most equal to, but generally lower than, his installed capacity/connected load.

6.8.5 **Gas appliance; gas-fired equipment; gas-fired installation** A common term for flued and flueless gas-burning appliances.

6.8.6 **Flueless gas appliance** An appliance designed for use without connection to a flue for venting the products of combustion to the exterior.

6.8.7 **Flued gas appliance** An appliance designed for use with connection to a flue for venting the products of combustion to the exterior.

6.8.8 **Flue; flue gas installation** The equipment required for venting to atmosphere the products of combustion from flued gas appliances.

6.8 Utilisation du Gaz

6.8.1 **Branchement** Ensemble comprenant les tuyauteries comprises entre la canalisation de distribution et le dispositif principal de coupure, le joint isolant et le dispositif principal de coupure lui-même, ainsi que le dispositif de coupure existant éventuellement en-dehors du bâtiment et le détendeur-régulateur d'immeuble.

6.8.2 **Conduite de branchement** Partie de tuyauterie comprise entre la canalisation de distribution et le dispositif principal de coupure.

6.8.3 **Installation d'abonné** Installation de gaz qui reçoit le gaz d'une distribution publique, considérée à partir du dispositif principal de coupure.

6.8.4 **Puissance installée** Quantité de gaz qui doit être fournie pour permettre d'atteindre le débit calorifique nominal. La puissance installée chez un abonné est la somme des puissances de tous les appareils d'utilisation du gaz qui sont raccordés à son installation.

N.B. La "puissance souscrite" est, en général, inférieure ou, au plus, égale à la puissance installée.

6.8.5 **Appareil d'utilisation (du gaz)** Terme général qui désigne des appareils d'utilisation du gaz non raccordés ou raccordés à un conduit de fumées.

6.8.6 **Appareil d'utilisation (du gaz) non raccordé** Appareil dont les produits de combustion ne sont pas évacués à l'air libre par un conduit de fumées.

6.8.7 **Appareil d'utilisation (du gaz) raccordé** Appareil d'utilisation du gaz dont les produits de combustion sont évacués à l'air libre par un conduit de fumées.

6.8.8 **Conduit de fumées** Dispositif qui permet d'évacuer dans l'atmosphère les produits de combustion des appareils d'utilisation du gaz raccordés.

6.8 Gasanwendung

6.8.1 **Hausanschluss** Hausanschluss ist die Gesamtheit von Hausanschlussleitungen, Isolierstück und Hauptabsperreinrichtung sowie ggf. Absperreinrichtung ausserhalb des Gebäudes und Hausdruckregelgerät.

6.8.2 **Hausanschlussleitung** Hausanschlussleitung ist das Leitungsteil zwischen Versorgungsleitung und Hauptabsperreinrichtung.

6.8.3 **Abnehmeranlage (Inneninstallation)** Abnehmeranlage ist eine installierte Gasanlage, die Gase der öffentlichen Versorgung abnimmt, ab Hauptabsperreinrichtung.

6.8.4 **Anschlusswert** Anschlusswert ist die Gasmenge, die zur Erreichung der Nennwärmebelastung zugeführt werden muss. Der Anschlusswert einer Abnehmeranlage ist die Summe der Nennleistungen aller angeschlossenen Gasverbrauchseinrichtungen.

Nota : Die Vertragsleistung eines Abnehmers ist in der Regel kleiner, höchstens aber gleich dem Anschlusswert.

6.8.5 **Gasverbrauchseinrichtung** ist der Sammelbegriff für Gasgeräte und Gasfeuerstätten.

6.8.6 **Gasgerät** Gasgerät ist eine Gasverbrauchseinrichtung, deren Abgase nicht durch eine Abgasanlage ins Freie abgeführt werden.

6.8.7 **Gasfeuerstätte** Gasfeuerstätte ist eine Gasverbrauchseinrichtung, deren Abgase über eine Abasanlage ins Freie abgeleitet werden.

6.8.8 **Abgasanlage** Abgasanlage ist eine Einrichtung zur Abführung der Abgase von Gasfeuerstätten ins Freie.

6.8 Utilización del Gas

6.8.1 **Acometida** Conjunto que comprende las tuberías entre la canalización de distribución y la llave general de paso, la junta aislante y la misma llave general de paso así como la llave de paso que puede existir eventualmente por fuera del edificio y el regulador de presión del inmueble.

6.8.2 **Tubo de acometida** Parte de tubería comprendida entre la canalización de distribución y la llave general de paso.

6.8.3 **Instalación del abonado** Instalación de gas que, a partir de la salida de la llave general de paso, recibe el gas de la red de suministro.

6.8.4 **Capacidad instalada** Cantidad de gas que debe suministrarse para permitir alcanzar el caudal calorífico normal. La capacidad instalada en el domicilio de un abonado es la suma de las potencias de todos los aparatos que utilizan el gas que están conectados a su instalación.

Nota. La capacidad contratada es, en general, inferior o, a lo sumo, igual a la capacidad instalada.

6.8.5 **Aparato de utilización (del gas)** Expresión comúnmente utilizada para designar tanto los aparatos conectados como los no conectados a los conductos de evacuación de humos.

6.8.6 **Aparato no conectado a la evacuación de humos** Aparato para utilización del gas no conectado al conducto de evacuación de humos.

6.8.7 **Aparato conectado a la evacuación de humos** Aparato para utilizar el gas cuyos productos de combustión se evacuan al aire libre a través de un conducto de evacuación.

6.8.8 **Evacuación de humos** Dispositivo que permite evacuar a la atmósfera los productos de la combustión de los aparatos para utilización del gas conectados a la misma.

6

Section 7

Nuclear Power Technology
Industrie nucléaire Energie nucléaire
Kernenergetik Kernenergiewirtschaft
Industria Nuclear Energia Nuclear

7

Nuclear Power Technology

7.1 Basic Terms

7.1.1 Nuclear power station A power station that employs one or more power reactors to generate electric or thermal energy.

7.1.2 Nuclear reactor A device in which a self-sustaining nuclear fission chain reaction can be maintained and controlled (fission reactor). The term is sometimes applied to a device in which a nuclear fusion reaction can be produced and controlled (fusion reactor). (Also called *reactor* or *pile*.)

7.1.3 Power reactor A reactor whose primary purpose is to produce energy. Reactors in this class include: (a) electric power reactors; (b) process-heat reactors; and (c) propulsion reactors.

7.1.4 Thermal reactor A reactor in which fission is induced predominantly by thermal neutrons.

7.1.5 Homogeneous reactor A reactor in which the core materials are distributed in such a manner that its neutron characteristics can be accurately described by the assumption of a homogeneous distribution of the materials throughout the core.

7.1.6 Heterogeneous reactor A reactor in which the core materials are segregated to such an extent that its neutron characteristics cannot be accurately described by the assumption of homogeneous distribution of the materials throughout the core.

7.1.7 Fast reactor A reactor in which fission is induced predominantly by fast neutrons. Also called *fast neutron reactor.*

7.1.8 Breeder reactor; breeder A reactor which produces more fissile material than it consumes, i.e. has a conversion ratio greater than unity.

7.1.9 Nuclear fuel A substance containing one or more fissile nuclides capable of maintaining a chain reaction in a reactor; also a substance containing one or more fertile nuclides that can be transmuted into such fissile nuclides.

7.1.10 Fission products Nuclides produced either by fission or by the subsequent radioactive decay of the nuclides thus formed.

7.1.11 Radioactivity The property of certain nuclides of spontaneously emitting particles, including gamma radiation, from their nucleus, of undergoing spontaneous fission or of emitting X radiation following orbital electron capture by their nucleus.

7.1.12 Source material; feed material (UK) Uranium containing the mixture of isotopes occurring in nature; uranium depleted in the isotope 235; thorium; any of the foregoing in the form of metal, alloy, chemical compound, or concentrate.

Industrie nucléaire
Energie nucléaire

7.1 Notions Fondamentales

7.1.1 Centrale nucléaire Centrale par laquelle l'énergie électrique ou thermique est produite par un ou plusieurs réacteurs de puissance.

7.1.2 Réacteur nucléaire - réacteur Dispositif dans lequel une réaction de fission nucléaire en chaîne auto-entretenue peut être maintenue et dirigée. Ce terme est quelquefois appliqué à un dispositif dans lequel une réaction de fusion nucléaire peut être produite et dirigée.

7.1.3 Réacteur de puissance Réacteur conçu principalement pour produire de l'énergie. Parmi les réacteurs de puissance on distingue:
— les réacteurs de production d'électricité
— les réacteurs de production de chaleur
— les réacteurs de propulsion

7.1.4 Réacteur thermique Réacteur dans lequel la fission est produite principalement par des neutres thermiques.

7.1.5 Réacteur homogène Réacteur dans lequel les matériaux du coeur sont répartis de telle sorte que ses caractéristiques neutroniques peuvent être convenablement décrites avec une hypothèse homogène de ces matériaux dans le coeur.

7.1.6 Réacteur hétérogène Réacteur dans lequel les matériaux du coeur ne sont pas mélangés intimement les uns aux autres, de telle sorte que les caractéristiques neutroniques sont influencées par la structure du mélange.

7.1.7 Réacteur rapide Réacteur dans lequel la fission est produite principalement par des neutrons rapides.
Synonyme ''réacteur à neutrons rapides''.

7.1.8 Réacteur surrégénérateur Réacteur produisant plus de matière fissile qu'il n'en consomme, c'est-à-dire ayant un rapport de conversion plus grand que 1.

7.1.9 Combustible nucléaire Matériau contenant des nucléides fissiles capables d'entretenir une réaction nucléaire en chaîne dans un réacteur (y compris également les nucléides fertiles).

7.1.10 Produits de fission Nucléides produits soit par fission, soit par la désintégration radioactive ultérieure des nucléides formés de cette façon.

7.1.11 Radioactivité Propriété qu'ont certains nucléides d'émettre spontanément des particules ou un rayonnement gamma, à partir de leur noyau, de se scinder spontanément, ou d'émettre un rayonnement X après capture d'un électron orbital par le noyau.

7.1.12 Substances de base La notion de substance de base est utilisée dans les documents internationaux pour désigner les matières Suivantes:
— uranium, à mélange isotopique naturel,
— uranium, à teneur en U 235 inférieure à la teneur naturelle,
— thorium,
chacune des substances ci-dessus se présentant sous forme de métal, d'alliage, de combinaison chimique ou de concentré.

Kernenergetik
Kernenergiewirtschaft

7.1 Grundbegriffe

7.1.1 Kernkraftwerk Ein Kernkraftwerk ist ein Kraftwerk, in dem elektrische Energie oder Wärmeenergie mit Hilfe eines Leistungsreaktors oder mehrerer Leistungsreaktoren erzeugt wird.

7.1.2 Kernreaktor; Reaktor Ein Kernreaktor ist eine Einrichtung, in der eine sich selbst erhaltende Kettenreaktion von Kernspaltungen aufrechterhalten und gesteuert werden kann. Dieser Begriff kann auch für eine Anlage angewendet werden, die der Fusion dient.

7.1.3 Leistungsreaktor Ein Leistungsreaktor ist ein Reaktor, der vorwiegend der Energieerzeugung dient. Zu den Leistungsreaktoren gehören :
Stromerzeugungsreaktoren
Wärmeerzeugungsreaktoren
Antriebsreaktoren

7.1.4 Thermischer Reaktor Ein thermischer Reaktor ist ein Reaktor, bei dem die Spaltungen vorwiegend von thermischen Neutronen ausgelöst werden.

7.1.5 Homogener Reaktor Ein homogener Reaktor ist ein Reaktor, in dem die Materialien der Spaltzone so innig miteinander vermischt sind, dass das Verhalten der Neutronen von der Struktur der Mischung nicht beeinflusst wird.

7.1.6 Heterogener Reaktor Ein Reaktor, in dem die Materialien der Spaltzone nicht innig miteinander vermischt sind, so dass das Verhalten der Neutronen von der Struktur der Mischung beeinflusst wird.

7.1.7 Schneller Reaktor Ein schneller Reaktor ist ein Reaktor, bei dem die Spaltungen vorwiegend von schnellen Neutronen ausgelöst werden.
Auch : Schnellneutronenreaktor

7.1.8 Brutreaktor ; Brüter Ein Brutreaktor ist ein Reaktor, der mehr Spaltstoff erzeugt als er verbraucht, d.h. mit einem Konversionsverhältnis grösser als eins arbeitet.

7.1.9 Kernbrennstoff; Brennstoff Kernbrennstoff ist Material, das einen. oder mehrere Spaltstoffe enthält, welche eine Kettenreaktion aufrechterhalten können (einschliesslich Brutstoffe).

7.1.10 Spaltprodukte Spaltprodukte sind Nuklide, die durch Spaltung oder radioaktive Umwandlung der durch Spaltung gebildeten Nuklide entstehen.

7.1.11 Radioaktivität Unter Radioaktivität versteht man die Eigenschaft bestimmter Nuklide, spontan Teilchen einschliesslich Gammaquanten aus ihrem Kern zu emittieren, sich spontan zu spalten oder - nach Einfang eines Hüllenelektrons durch den Kern - Röntgenstrahlung aus der Hülle auszusenden.

7.1.12 Ausgansstoffe; Ausgangsmaterial Der Begriff Ausgangsstoffe wird in internationalen Dokumenten zur Bezeichnung folgender Stoffe benutzt : Uran, das die in der Natur auftrende Isotopenmischung enthält;
Uran, dessen Gehalt an Uran 235 unter dem natürlichen Gehalt liegt;
Throium;
Jeder der erwähnten Stoffe in Form von Metall, Legierung, chemischer Verbindung oder von Konzentrat.

Industria Nuclear
Energia Nuclear

7.1 Conceptos Fundamentales

7.1.1 Central nuclear Central eléctrica en la que la energía eléctrica o térmica se produce mediante uno o varios reactores de potencia.

7.1.2 Reactor nuclear; reactor Dispositivo en el que se puede sostener y controlar una reacción automantenida de fisión nuclear en cadena. Se aplica algunas veces este termino a un dispositivo en el que se puede producir y controlar una reacción de fusión termonuclear.

7.1.3 Reactor de potencia Reactor especialmente disenado para producir energía. Son reactores de este tipo los de producción de electricidad, producción de calor y propulsión.

7.1.4 Reactor térmico Reactor en el que la fisión se produce principalmente por neutrones térmicos.

7.1.5 Reactor homogéneo Reactor en el que los materiales del núcleo se distribuyen de tal forma, que las características neutrónicas pueden ser convenientemente descritas con una hipótesis de reparto homogéneo de estos materiales en el núcleo del reactor.

7.1.6 Reactor heterogéneo Reactor en el que los materiales del núcleo se distribuyen de tal forma que las características neutrónicas no pueden ser convenientement descritas con una hipótesis de reparto homogéneo de estos materiales en el núcleo del reactor.

7.1.7 Reactor rápido Reactor en el que la fisión es producida principalmente por neutrones rápidos.
Sinónimo: reactor de neutrones rápidos.

7.1.8 Reactor reproductor Reactor que produce más materia fisible que la que consume, es decir, que posee un factor de conversión mayor que la unidad.

7.1.9 Combustible nuclear Material que contiene nucleidos fisibles capaces de mantener una reacción nuclear en cadena en un reactor (incluidos igualmente los nucleidos fértiles).

7.1.10 Productos de fisión Nucleidos producidos bien por fisión, bien por desintegración radiactiva ulterior, de los nucleidos formados de ests manera.

7.1.11 Radiactividad Propiedad de ciertos nucleidos de emitir espontáneamente partículas o radiación gamma, o de emitir radiación x tras la captura de un electrón orbital o de sufrir una fisión espontánea.

7.1.12 Material básico Se utiliza este término en los documentos internacionales para designar el formado por una mezcla isotópica natural de uranio; por una mezcla cuyo contenido de uranio 235 es menor que el natural, y torio; y cualquiera de los elementos citados en forma de metal, aleacion, mezcla quimica o concentrado.

7

7.1.13 Special fissionable material; special nuclear material Plutonium-239; uranium-233; uranium enriched in the isotopes 235 or 233; any material containing one or more of the foregoing; but the term "special fissionable material" does not include source material.

Note. The term "uranium enriched in the isotopes 235 or 233" means uranium containing the isotopes 235 or 233 or both in an amount such that the abundance ratio of the sum of these isotopes to the isotope 238 is greater than the ratio of the isotope 235 to the isotope 238 occurring in nature.

7.1.14 Fuel inventory The total amount of nuclear fuel invested in a reactor, a group of reactors, or an entire fuel cycle, according to specification.

7.1.15 Fissile material inventory The total amount of fissile material invested in a reactor, a group of reactors or an entire fuel cycle, according to specification.

7.1.16 Fuel cycle The sequence of steps, such as utilisation, reprocessing, refabrication and eventual re-utilisation in reactors, through which nuclear fuel may pass.

7.1.17 Nuclear fission The division of a heavy nucleus into two (or, rarely, more) parts with masses of equal order of magnitude, usually accompanied by the emission of neutrons and gamma radiation.

7.1.18 Fission energy The energy released when an atom is split.

7.1.19 Thermal neutrons Neutrons in thermal equilibrium with the medium in which they exist.

7.1.20 Fast neutrons Neutrons of kinetic energy greater than some specified value. In reactor physics the value is frequently chosen to be 0.1 MeV.

7.1.21 Fission neutrons Prompt and delayed neutrons originating in the fission process that have retained their original energy.

7.1.22 Prompt neutrons Neutrons accompanying the fission process without measurable delay.

7.1.23 Delayed neutrons Neutrons resulting from fission which are emitted with a measurable time delay due to the radioactive disintegration of fission products or which are formed late such as photo-neutrons.

7.1.24 Criticality The condition of being critical.

7.1.25 Nuclear chain reaction A series of nuclear reactions in which one of the agents necessary to the series is itself produced by the same reactions. Depending on whether the number of reactions directly caused by one reaction is on average less than, equal to, or greater than, unity, the chain reaction is convergent (subcritical), self-sustained (critical), or divergent (supercritical).

7.1.26 Moderation The reduction of the neutron energy by scattering without appreciable capture.

7.1.13 Matières fissiles, spéciales - matériau fissile, spécial Sont considérées comme fissiles, dans les documents internationaux, les matiéres suivantes:
— plutonium 239, uranium 233, uranium enrichi en isotopes 235 ou 233, et toute matière contenant une ou plusieurs des matières précitées. Les matières fissiles spéciales n'englobent toutefois pas les produits de base

Remarque : l'expression "uranium enrichi en isotopes 235 (ou 233)" désigne l'uranium contenant les isotopes 235 (ou 233) ou les deux, en quantité telle que le rapport de la somme de ces deux isotopes à l'isotope 238 est supérieur au rapport existant à l'état naturel entre l'isotope 235 et l'isotope 238.

7.1.14 Inventaire de combustible Quantité totale du combustible nucléaire investi dans un réacteur, un ensemble de réacteurs ou un cycle de combustible tout entier.

7.1.15 Inventaire de matière fissile Quantité de matière fissile placée dans un réacteur, un ensemble de réacteurs ou un cycle de combustible tout entier.

7.1.16 Cycle du combustible Succession des étapes que parcourt le combustible nucléaire telles que l'usage du combustible nucléaire dans le réacteur, son retraitement après combustion et sa réutilisation éventuelle dans les réacteurs.

7.1.17 Fission nucléaire, fission Division d'un noyau lourd, généralement en deux parties (ou rarement plus) dont les masses sont du même ordre de grandeur, habituellement accompagnée de l'émission de neutrons et de rayonnement gamma.

7.1.18 Energie de fission Energie libérée par la fission d'un atome.

7.1.19 Neutrons thermiques Neutrons en équilibre thermique avec le milieu dans lequel ils se trouvent.

7.1.20 Neutrons rapides Neutrons d'énergie cinétique supérieurs à une certaine valeur spécifiée. En physique des réacteurs, cette valeur est souvent fixée à 0,1 MeV.

7.1.21 Neutrons de fission Neutrons accompagnant immédiatement ou avec retard le processus de fission et qui ont gardé leur énergie initiale.

7.1.22 Neutrons instantanés Neutrons accompagnant le processus de fission sans retard mesurable.

7.1.23 Neutrons différés (neutrons retardés) Neutrons qui ne sont pas libérés immédiatement par le processus de fission nucléaire, mais avec un certain retard dû à une désintégration radioactive de produits de fission ou qui se forment avec retard comme photoneutrons.

7.1.24 Criticité Etat de ce qui est critique.

7.1.25 Réaction en chaîne Série de réactions nucléaires, dans lesquelles l'un des agents nécessaires à la série est lui-même produit par les mêmes réactions.

7.1.26 Modération Réduction de l'énergie cinétique des neutrons par diffusion sans capture appréciable.

7.1.13 Spaltbare Stoffe, besondere; Spaltbares Material, besonderes Als besondere spaltbare Stoffe werden in internationalen Dokumenten folgende Stoffe bezeichnet : Plutonium 239; Uran 233; mit den Isotopen 235 oder 233 angereichertes Uran und jeder Stoff, der einen oder mehrere der vorerwähnten Stoffe enthält. Besondere spaltbare Stoffe schliessen jedoch Ausgangsstoffe nicht ein.

Anmerkung : Der Ausdruck "mit den Isotopen 235 oder 233 angereichertes Uran" bedeutet Uran, das die Isotope 235 oder 233 oder diese beiden Isotope in einer solchen Menge enthält, dass das Verhältnis der Summe dieser beiden Isotope zum Isotop 238 grösser ist als das in der Natur auftretende Verhältnis des Isotops 235 zum Isotop 238.

7.1.14 Brennstoffeinsatz; Brennstoffinventar Der Brennstoffeinsatz ist die Gesamtmenge Kernbrennstoff, die in einen Reaktor, ein Reaktorsystem oder einen ganzen Brennstoffkreislauf eingesetzt wird, je nach Spezifizierung.

7.1.15 Spaltstoffeinsatz; Spaltstoffinventar Der Spaltstoffeinsatz ist die Gesamtmenge Spaltstoff, die in einem Reaktor, ein Reaktorsystem oder einem ganzen Brennstoffkreislauf eingesetzt wird, je nach Spezifizierung.

7.1.16 Brennstoffkreislauf Als Brennstoffkreislauf bezeichnet man die vom Kernbrennstoff durchlaufenen Etappen, wie Nutzung des Kernbrennstoffs im Reaktor, Aufarbeitung nach Abbrand und etwaige Wiederverwendung in Reaktoren.

7.1.17 Kernspaltung; Spaltung Die Kernspaltung ist die Teilung eines schweren Kerns meistens in zwei mittelschwere Kerne und sehr viel seltener in mehrere Kerne. Der Vorgang ist mit der Emission von Neutronen und Gammastrahlen verbunden.

7.1.18 Spaltungsenergie Spaltungsenergie ist die bei der Spaltung eines Atoms freigesetzte Energie.

7.1.19 Thermische Neutronen Thermische Neutronen sind Neutronen, die sich im thermischen Gleichgewicht mit dem umgebenden Medium befinden.

7.1.20 Schnelle Neutronen Schnelle Neutronen sind Neutronen, die kinetische Energien oberhalb eines bestimmten Wertes haben. In der Reaktorphysik wird häufig ein Wert von etwa 0,1 MeV gewählt.

7.1.21 Spaltneutronen Spaltneutronen sind die bei der Kernspaltung prompt und verzögert entstehenden Neutronen, die ihre ursprüngliche Energie behalten haben.

7.1.22 Prompte Neutronen Prompte Neutronen sind Neutronen, die beim Spaltprozess ohne messbare Verzögerung freigesetzt werden.

7.1.23 Verzögerte Neutronen Verzögerte Neutronen sind Neutronen, die bei einer Kernspaltung nicht unmittelbar, sondern als Folge einer radioaktiven Umwandlung von Spaltprodukten oder als Photoneutronen verspätet entstehen.

7.1.24 Kritikalität Unter Kritikalität versteht man den Zustand eines Systems, das kritisch ist.

7.1.25 Ketenreaktion Die Kettenreaktion ist eine Folge von Kernreaktionen, deren jede von den Produkten vorhergehender Kernreaktionen ausgelöst wird.

7.1.26 Moderation Die Moderation ist der Vorgang, bei dem die kinetische Energie der Neutronen durch Streustösse vermindert wird, ohne dass übermässige Neutronenverluste durch Absorption entstehen.

7.1.13 Materiales fisibles, especiales; material fisible especial Se consideran como fisibles, en los documentos internacionales los materiales siguientes:

Plutonio 239, uranio 233, uranio enriquecido en isótopos 235 o 233, y todo material conteniendo uno o varios de los materiales precitados. Los materiales fisibles especiales no comprenden siempre los materiales básicos.

Observación: La expresión "uranio enriquecido en isótopos 235 o 233 "designa el uranio conteniendo los isótopos 235 o 233 o ambos, en cantidad tal, que la relación de la suma de estos dos isótopos al isótopo 238 es superior a la relación existente en el estado natural entre el isótopo 235 y el isótopo 238.

7.1.14 Dotación de combustible La cantidad total de combustible nuclear que se utiliza en un reactor, un grupo de reactores, o un ciclo completo de combustible.

7.1.15 Dotación de material fisible Cantidad total de material fisible que se utiliza en un reactor, un grupo de reactores o en un ciclo completo de combustible.

7.1.16 Ciclo de combustible Sucesión de etapas que recorre el combustible nuclear, tales como la utilización del combustible nuclear en el reactor, su reprocesado después de la combustión, y su posible reutilización en los reactores.

7.1.17 Fisión nuclear División de un núcleo pesado en dos partes (raramente en más) cuyas masas son del mismo orden de magnitud, acompañada habitualmente de emisión de neutrones y de radiaciones gamma.

7.1.18 Energía de fisión Energía liberada por la fisión de un átomo.

7.1.19 Neutrones térmicos Neutrones en equilibrio térmico con el medio en que se encuentran.

7.1.20 Neutrones rápidos Neutrones de energía cinética superior a un cierto valor especificado. En la física de reactores este valor se fija frecuentemente en 0,1 MeV.

7.1.21 Neutrones de fisión Neutrones producidos inmediatamente o con retardo en los procesos de fisión y conservando su energía inicial.

7.1.22 Neutrones inmediatos Neutrones que accompañan el proceso de fisión sin retardo medible.

7.1.23 Neutrones diferidos (neutrones retardados) Neutrones que no son liberados inmediatamente por el proceso de fisión nuclear, sino con un cierto retardo debido a una desintegración radiactiva de productos de fisión o que se forman con retardo como fotoneutrones.

7.1.24 Criticidad La condición de ser crítico.

7.1.25 Reacción nuclear en cadena Serie de reacciones nucleares en las que uno de los agentes necesarios a la serie es producido, él mismo, por las propias reacciones.

7.1.26 Moderación Reducción de la energía cinética de los neutrones por difusión sin captura apreciable.

7

7.1.27 Critical mass The minimum mass of fissile material or of an element having a specified degree of enrichment in fissile material, with a specified geometrical arrangement, material composition and environment, that can sustain a critical chain reaction.

7.1.28 Multiplication factor; effective multiplication factor The ratio, k, of the total number of neutrons produced during a time interval to the total number of neutrons lost by absorption and leakage during the same interval.

Note. The "multiplication factor" is also termed the "effective multiplication factor" to differentiate it from the "infinite multiplication factor".

7.1.29 Critical A medium is critical when it has an effective multiplication factor equal to unity.

7.1.30 Supercritical A medium is supercritical when its effective multiplication factor is greater than unity.

7.1.31 Subcritical A medium is subcritical when its effective multiplication factor is less than unity.

7.1.32 Breeding ratio The conversion ratio when it is greater than unity.

7.1.33 Conversion ratio The ratio of the number of fissile nuclei produced from fertile material to the number of fissile nuclei consumed during the same period of time.

7.1.34 Doubling time
1. For the fuel inventory of an entire fuel cycle of a breeder reactor or group of breeder reactors, the time required for the amount of fissile nuclides to be doubled by breeding;
2. For a fuel charge in a given breeder reactor, the time required for the initial amount of fissile nuclides to be doubled by breeding.

7.1.35 Reactivity The parameter, ρ, giving the deviation from criticality of a nuclear-chain-reacting medium. It is defined as follows:

$$\rho = \frac{k-1}{k}$$

where ρ = reactivity and k = multiplication factor.

7.1.36 Nuclear safety The prevention of plant conditions or operating conditions that could lead to the endangering of persons or objects by radioactive contamination, ionizing radiation or other release of energy.

7.2 Nuclear Fuels, Manufacture and Reprocessing

7.2.1 Natural uranium Uranium with the naturally occurring mixture of isotopes.

7.1.27 Masse critique Masse minimale de matière fissile ou d'un élément à un degré d'enrichissement déterminé en matières fissiles, qui pour une disposition géométrique, une composition matérielle et un environnement déterminés, peut permettre une réaction en chaîne auto-entretenue.

7.1.28 Facteur de multiplication, facteur effectif de multiplication Rapport k du nombre total de neutrons produits, au cours d'un intervalle de temps donné, au nombre total de neutrons perdus par absorption et par fuite au cours du même intervalle.

Remarque : par différence avec la notion de facteur de multiplication infini, le facteur de multiplication est aussi appelé facteur de multiplication effectif.

7.1.29 Critique Un milieu est dit critique lorsque le facteur de multiplication effectif est égal à 1.

7.1.30 Surcritique On appelle surcritique l'état d'un milieu dont le facteur de multiplication effectif est supérieur à 1.

7.1.31 Sous-critique On appelle sous-critique l'état d'un milieu dont le facteur de multiplication effectif est inférieur à 1.

7.1.32 Rapport de surrégénération Appellation donnée au rapport de conversion lorsqu'il est supérieur à 1.

7.1.33 Rapport de conversion (technique des réacteurs) Rapport du nombre de noyaux de matières fissiles produits par un élément fertile, en un temps donné, au nombre de noyaux de matières fissiles détruits pendant le même laps de temps.

7.1.34 Temps de doublement
1. En ce qui concerne le cycle du combustible : pour l'inventaire du combustible d'un cycle du combustible tout entier, le temps de doublement d'un réacteur surrégénérateur ou d'un ensemble de réacteurs surrégénérateurs est le temps nécessaire pour que le nombre de nucléides fissiles ait doublé par surrégénération.
2. En ce qui concerne le réacteur : pour une charge de combustible dans un réacteur surrégénérateur donné, le temps de doublement d'un réacteur surrégénérateur est le temps nécessaire pour que la quantité initiale de nucléides fissiles ait doublé par surrégénération.

7.1.35 Réactivité La réactivité d'un réacteur désigne l'écart qui le sépare de la criticité. Elle se définit par l'équation :

$$\rho = \frac{k-1}{k}$$

où ρ = réactivité
k = facteur de multiplication effectif

7.1.36 Sûreté nucléaire Dans les installations nucléaires, on désigne par sûreté nucléaire, les mesures destinées à empêcher l'apparition de situations ou de conditions de fonctionnement pouvant menacer la sécurité des personnes ou du matériel par contamination radioactive, rayonnement ionisant ou par toute autre énergie libérée.

7.2 Combustibles Nucléaires, Production et Retraitement du Combustible

7.2.1 Uranium naturel Uranium dont la composition isotopique est naturelle.

7.1.27 Kritische Masse Die kritische Masse ist die kleinste Masse eines bestimmten Spaltstoffs oder eines Elementes mit einer bestimmten Spaltstoffanreicherung in einer multiplizierenden Anordnung gegebener Form, Struktur, Grösse, Materialzusammensetzung und Umgebung, mit der in dieser eine sich selbst erhaltende Kettenraktion möglich ist.

7.1.28 Multiplikationsfaktor, effektiver Multiplikationsfaktor Der Multiplikationsfaktor ist das Verhältnis k zwischen der Gesamtzahl der Neutronen, die in einer bestimmten Zeitspanne entstehen, zu der Gesamtanzahl der Neutronen, die in der gleichen Zeitspanne in dem Medium absorbiert werden bzw. durch Ausströmen verlorengehen.

Anmerkung : Zur Abgrenzung gegenüber dem unendlichen Multiplikationsfaktor wird der Multiplikationsfaktor auch als effektiver Multiplikationsfaktor bezeichnet.

7.1.29 Kritisch Eine Anordnung heisst kritisch, wenn der effektive Multiplikationsfaktor eins ist.

7.1.30 Ueberkritisch Der Zustand einer Anordnung, deren effektiver Multiplikationsfaktor grösser als eins ist, heisst überkritisch.

7.1.31 Unterkritisch Der Zustand einer Anordnung, deren effektiver Multiplikationsfaktor kleiner als eins ist, heisst unterkritisch.

7.1.32 Brutverhältnis Das Brutverhältnis ist das Konversionsverhältnis, wenn dessen Wert grösser als eins ist.

7.1.33 Konversionsverhältnis (Reaktortechnik) Das Konversionsverhältnis ist das Verhältnis der Anzahl der aus einem Brutstoff erzeugten Kerne von Spaltstoffen zur Anzahl der während der gleichen Zeitspanne verbrauchten Kerne der vorhandenen Spaltstoffe.

7.1.34 Verdoppelungszeit
1. Die Verdoppelungszeit bezogen auf den Brennstoffkreislauf ist für einen Brutreaktor oder ein System von Brutreaktoren die Zeit, die zum Verdoppeln der Anzahl der Spaltstoffkerne im Spaltstoffeinsatz des Brennstoffkreislaufs durch Brüten benötigt wird.
2. Die Verdoppelungszeit bezogen auf den Reaktor ist für einen Brutreaktor die Zeit, die zum Verdoppeln der Anzahl der Spaltstoffkerne im Spaltstoffeinsatz des Reaktors durch Brüten benötigt wird.

7.1.35 Reaktivität Die Reaktivität eines Reaktors ist ein Mass für die Abweichung vom kritischen Zustand. Sie wird durch folgende Gleichung definiert :

$$\rho = \frac{k - 1}{k} \qquad \begin{array}{l} \rho = \text{Reaktivität} \\ k = \text{Multiplikationsfaktor} \end{array}$$

7.1.36 Nukleare Sicherheit Unter nuklearer Sicherheit versteht man in kerntechnischen Anlagen die Verhinderung von Anlagen- oder Betriebszuständen, die zu einer Gefährdung von Menschen oder Sachen durch radioaktive Kontamination, ionisierende Strahlung oder andere freigesetzte Energie führen können.

7.2 Nukleare Brennstoffe, Brennstoffherstellung und -aufarbeitung

7.2.1 Natururan Natururan ist Uran mit der natürlichen Isotopenzusammensetzung.

7.1.27 Masa crítica Masa mínima de material fisible o de un elemento con un grado de enriquecimiento determinado en materiales fisibles, que puede permitir una reacción en cadena automantenida, si se dan una disposición geométrica, una composición material y un entorno determinados.

7.1.28 Factor de multiplicación Relación del número total de neutrones producidos, durante un intervalo de tiempo, al número total de neutrones perdidos por absorción y fuga, durante el mismo intervalo de tiempo. Cuando se evalúa esta cantidad para un medio infinito o una red que se repite indefinidamente se designa este factor como factor de multiplicación en un medio infinito (k s); cuando se evalua para un medio finito se designa como factor efectivo de multiplicación (kef).

7.1.29 Crítico Dícese de un sistema nuclear que tiene un factor de multiplicación efectivo igual a la unidad.

7.1.30 Supercrítico Se denomina supercrítico el estado de un sistema cuyo factor de multiplicación es mayor que 1.

7.1.31 Subcrítico Se denomina subcrítico el estado de un sistema cuyo factor de multiplicación es menor que 1.

7.1.32 Razón de reproducción Nombre que se da al factor de conversión cuando es superior a la unidad.

7.1.33 Razón de conversión Relación del número de todos los núcleos fisibles producidos a partir de un mateial fértil, en un tiempo dado, al número de todos los núcleos fisibles detruidos en el mismo período de tiempo.

7.1.34 Tiempo de doblado
1. En lo que se refiere al ciclo de combustible: para el inventario del combustible de un ciclo completo, el tiempo de doblado de un reactor reproductor o de un conjunto de reactores reproductores es el tiempo necesario para que el número de nucleidos fisibles se haya doblado por reproducción.
2. En lo que se refiere al reactor para una carga de combustible en un determinado reactor reproductor, el tiempo necesario para que la cantidad inicial de nucleidos fisibles se haya doblado por reproducción.

7.1.35 Reactividad La reactividad de un reactor designa la desviación que le separa del estado crítico. Se define por la ecuación:

$$\rho = \frac{k-1}{k}$$

siendo ρ: Reactividad
k: Factor de multiplicación efectivo

7.1.36 Seguridad nuclear En las instalaciones nucleares se designa como seguridad nuclear las medidas destinadas a impedir la aparición de situaciones o de condiciones de funcionamiento que puedan amenazar la seguridad de las personas o de los materiales por contaminación reactiva, por radiación ionizante o por toda otra energía liberada.

7.2 Combustibles Nucleares, Producción y Reprocesado del Combustible

7.2.1 Uranio natural Uranio de composición isotópica natural.

7

7.2.2 Enriched uranium Uranium in which the percentage of the fissionable isotope, uranium-235, has been increased above that contained in natural uranium.

7.2.3 Fertile A nuclide is deemed to be fertile when it is capable of being transformed, directly or indirectly, into a fissile nuclide by neutron capture.

7.2.4 Fertile material Isotopes capable of being readily transformed, directly or indirectly, into fissionable material by the absorption of neutrons, particularly uranium-238 and thorium-232; sometimes called *source material* or *breeder material*.

7.2.5 Enrichment The fraction of atoms of a specified isotope in a mixture of isotopes of the same element when this fraction exceeds that in the naturally occurring mixture.

7.2.6 Enrichment process Any process by which the content of a specified isotope in an element is increased. The following are, inter alia, recognised uranium enrichment processes: gas diffusion, gas centrifuging, nozzle separation.

7.2.7 Enriched fuel Nuclear fuel containing uranium in which the uranium-233 and uranium-235 isotopes are present in an amount such that the abundance ratio of the sum of these isotopes to the isotope 238 is greater than the ratio of the isotope 235 to the isotope 238 occurring in nature, or to which chemically different fissile nuclides have been added.

7.2.8 Fuel reprocessing The processing of nuclear fuel after its use in a reactor, to remove fission products and recover fissile and fertile material.

7.2.9 Depleted uranium Uranium having less than the natural content of the easily fissionable uranium-235, e.g. the residue from an enrichment plant or some reactors.

7.2.10 Plutonium recovery The recovery of plutonium in the reprocessing of irradiated fuel.

7.2.11 Plutonium recycling The re-use of recovered plutonium in reactors.

7.2.12 Fissile material; fissionable material A nuclide readily fissioned by slow neutrons, for example, uranium-235, uranium-233, plutonium-239, plutonium-241.

7.3 Power Reactors, Main and Auxiliary Components

7.3.1 Pressure tube reactor A reactor whose fuel assemblies and coolant are confined in tubes that withstand the pressure of the coolant.

7.3.2 Boiling water reactor (BWR) A reactor in which water is used as coolant and moderator and allowed to boil in the core. In the case of a power reactor, the steam produced in the reactor vessel can be supplied directly to a turbo-alternator, but it will be slightly radioactive. It requires enriched fuel.

7.2.2 Uranium enrichi L'Uranium est dit enrichi lorsque sa teneur en uranium 235 est supérieur à ce qu'elle est à l'état naturel.

7.2.3 Fertile Un nucléide est dit fertile lorsqu'il est susceptible d'être transformé directement ou indirectement en une matière fissile, par capture de neutron.

7.2.4 Matière fertile Un nucléide est considéré comme fertile s'il est capable d'être transformé directement ou indirectement en nucléide fissile, par capture de neutrons.

Le thorium 232 et l'uranium 238 sont des matières fertiles.

7.2.5 Enrichissement Teneur isotopique relative à un isotope déterminé présent dans un élément lorsque cette teneur est supérieure à la teneur isotopique naturelle.

7.2.6 Procédés d'enrichissement Méthodes permettant d'augmenter la teneur d'un élément en un isotope déterminé. Les procédés d'enrichissement de l'uranium, entre autres, sont la diffusion gazeuse, l'ultracentrifugation et la séparation isotopique par tuyère.

7.2.7 Combustible enrichi Un combustible est dit enrichi lorsque la proportion de nucléides d'uranium 235 (et d'uranium 233) présents dans l'uranium qu'il contient est supérieure à celle d'uranium 235 dans l'uranium naturel, ou lorsque d'autres matières fissiles y ont été ajoutées.

7.2.8 Retraitement du combustible, traitement du combustible irradié Traitement du combustible nucléaire après son utilisation dans un réacteur, en vue d'extraire les produits de fission et de restaurer les matériaux fissiles et fertiles.

7.2.9 Uranium appauvri L'Uranium est dit appauvri lorsque sa teneur en uranium 235, seule matière fissile qu'il contienne, est inférieure à ce qu'elle est à l'état naturel.

7.2.10 Récupération du plutonium Extraction du plutonium contenu dans le combustible irradié par retraitement de celui-ci.

7.2.11 Recyclage du plutonium Réutilisation dans les réacteurs du plutonium récupéré.

7.2.12 Matière fissile Nucléide susceptible de subir une fission par interaction avec des neutrons lents.

Remarque : les matières fissiles présentant un intérêt pratique sont l'uranium 233, l'uranium 235, le plutonium 239 et le plutonium 241.

7.3 Réacteurs de Puissance, Constituants Essentiels et Installations Auxiliaires

7.3.1 Réacteur à tubes de force Réacteur dont les assemblages combustibles et le fluide de refroidissement sont enfermés dans des tubes qui supportent la pression du fluide de refroidissement.

7.3.2 Réacteur à eau bouillante (B.W.R.) Réacteur dans lequel l'eau servant de fluide de refroidissement et de modérateur peut bouillir. La vapeur est produite directement dans le caisson du réacteur et peut être fournie à la turbine, mais en étant légèrement radioactif. Ce type nécessite de l'uranium enrichi.

7.2.2 Angereichertes Uran Angereichertes Uran enthält einen grösseren Anteil an spaltbaren Uran 235 als im natürlichen Zustand.

7.2.3 Brütbar Als brütbar bezeichnet man Nuklide, die direkt oder indirekt durch Neutroneneinfang in Spaltstoffe umgewandelt werden können.

7.2.4 Brutstoff Brutstoffe sind Nuklide, die durch Neutronenbeschuss direkt oder indirekt in spaltbare Stoffe umgewandelt werden können, z.B. Thorium und Uran 238.

7.2.5 Anreicherung Anreicherung bezeichnet den erhöhten Gewichtsanteil eines bestimmten Isotops in einem Element gegenüber dem natürlichen Zustand.

7.2.6 Anreicherungsverfahren Anreicherungsverfahren sind Verfahren, mit dem der Anteil eines bestimmten Isotops in einem Element vergrössert wird. Anreicherungsverfahren für Uran sind unter anderem :

Gasdiffussionsverfahren
Gaszentrifugen-Verfahren
Trenndüsen-Verfahren

7.2.7 Angereicherter Brennstoff Angereicherter Brennstoff ist Kernbrennstoff, der Uran enthält, in dem die Nuklide Uran 233 und Uran 235 mit einer relativen Häufigkeit vorkommen, die über der relativen Häufigkeit von Uran 235 im Natururan liegt, oder dem andere Spaltstoffe zugefügt wurden.

7.2.8 Brennstoffaufarbeitung; Wiederaufarbeitung Brennstoffaufarbeitung ist die Aufarbeitung von Kernbrennstoff nach seiner Verwendung im Reaktor mit dem Ziel, die Spaltprodukte zu entfernen sowie Spalt- und Brutstoffe zurückzugewinnen.

7.2.9 Abgereichertes Uran, verarmtes Uran Abgereichertes Uran ist Uran, in dem das Nuklid Uran 235 als einziger Spaltstoff mit einer relativen Häufigkeit enthalten ist, die unter seiner relativen Häufigkeit im Natururan liegt.

7.2.10 Plutoniumrückgewinnung Plutoniumrückgewinnung ist die Gewinnung von Plutonium bei der Brennstoffaufarbeitung.

7.2.11 Plutoniumrückführung Plutoniumrückführung ist die Wiederverwendung des bei der Plutoniumrückgewinnung erhaltenen Plutoniums in Reaktoren.

7.2.12 Spaltstoff Als Spaltstoff bezeichnet man ein Nuklid, das durch Wechselwirkung mit langsamen Neutronen gespalten werden kann.

Anmerkung : Spaltstoffe von praktischer Bedeutung sind : Uran 233, Uran 235, Plutonium 239, Plutonium 241.

7.3 Leistungsreaktoren, Hauptkomponenten und Nebenanlagen

7.3.1 Druckröhrenreaktor; Druckrohrreaktor Ein Druckröhrenreaktor ist ein Reaktor, dessen Brennelementbündel in Rohren angeordnet sind, in denen sie mit dem unter Druck stehenden Kühlmittel gekühlt werden.

7.3.2 Siedewasserreaktor Ein Siedewasserreaktor ist ein Reaktor, aus dem die erzeugte Wärme unter Verdampfung des als Reaktorkühlmittel dienenden Wassers abgeführt wird, wobei der Dampf leicht radioaktiv ist.

Dieser Reaktortyp benötigt leicht angereichertes Uran.

7.2.2 Uranio enriquecido Uranio en el que el porcentaje de isótopo fisible, uranio 235, es superior al que contiene el uranio natural.

7.2.3 Fértil Se denomina fértil el nucleido que puede ser transformado directa o indirectamente en material fisible mediante la captura de neutrones.

7.2.4 Material fértil Un nucleido se considera fértil si es capaz de ser transformado, directa o indirectamente, en nucleido fisible, por captura de electrones.

El torio 232 y el uranio 238 son materiales fértiles.

7.2.5 Enriquecimiento Concentración isotópica relativa a un determinado isótopo presente en un elemento, cuando esta concentración es superior a la que tiene en estado natural.

7.2.6 Métodos de enriquecimiento Procedimiento que permite aumentar la concentración en un isótopo determinado de un elemento. Son métodos de enriquecimiento del uranio, entre otros, la difusión gaseosa, la ultracentrifugación y la separación isotópica mediante toberas.

7.2.7 Uranio enriquecido Se denomina enriquecido un combustible, cuando la proporción de nucleidos de uranio 233 y de uranio 235 presentes en el uranio que contiene, es superior a la proporción de uranio 235 en el uranio natural, o si le han sido agregados otros materiales fisibles.

7.2.8 Reprocesado del combustible Tratamiento del combustible nuclear tras su utilización en un reactor, para separar los productos de fisión y recuperar los materiales fisibles y fértiles.

7.2.9 Uranio empobrecido Se denomina empobrecido el uranio cuya concentración en uranio 235, única materia fisible que contiene, es inferior a la existente en estado natural.

7.2.10 Recuperación de plutonio Extracción del plutonio contenido en el combustible irradiado mediante el reprocesado de este.

7.2.11 Reciclado del plutonio Reutilización del plutonio recuperado en los reactores.

7.2.12 Material fisible Nucleido susceptible de sufrir una fisión por interacción con neutrones lentos.

Observación: Los materiales fisibles que presentan un interés práctico son el uranio 233, el uranio 235, el plutonio 239 y el plutonio 241.

7.3 Reactores de Potencia, Componentes Principales e Instalaciones Auxiliares

7.3.1 Reactor de tubos de presión Reactor cuyo conjunto combustible y refrigerante están contenidos en tubos que soportan la presión del fluido refrigerante.

7.3.2 Reactor de agua en ebullición (B.W.R.) Reactor en el que el agua que se usa como refrigerante y moderador puede estar en ebullición en el núcleo. Se produce vapor directamente en la vasija de seguridad, y en este estado puede ser suministrado a una turbina, aunque sea ligeramente radiactivo. Este tipo de reactor requiere uranio enriquecido.

7

7.3.3 Pressurised water reactor (PWR) A reactor in which the water coolant and moderator is kept at a high pressure to prevent it readily boiling and hence to keep it liquid. In the case of a power reactor, steam produced by heat exchange with the coolant is supplied to a turbo-alternator. It requires enriched fuel.

7.3.4 Gas-cooled reactor (GCR) A reactor in which gas is used as coolant and graphite as moderator. In the case of a power reactor, steam produced by heat exchange with the coolant gas is supplied to a turbo-alternator. The gas-cooled reactor, sometimes referred to as the Magnox type, uses natural uranium; the Advanced gas-cooled reactor (AGR) and the High-temperature gas-cooled reactor (HTGR) require enriched fuel.

7.3.5 Heavy-water reactor (HWR) A reactor that uses heavy water as its moderator. The coolant may be gas as in the ''Heavy-water-moderated, gas-cooled reactor'' (HWGCR), light water as in the ''Heavy-water-moderated, boiling light-water-cooled reactor'' (HWLWR) or ''Steam-generating heavy-water reactor'' (SGHWR), or heavy-water as in the ''Pressurised heavy-water-moderated and cooled reactor'' (PHWR). In the case of a power reactor, steam produced in the reactor vessel or by heat exchange with the coolant is supplied to a turbo-generator. According to the type of plant, natural uranium or enriched fuel are required.

7.3.6 High-temperature reactor (HTR) (HTGR) A reactor that by the use of noble gases as reactor coolant and by the use of ceramic materials in the reactor core can be operated with high coolant exit temperatures. Graphite is employed as moderator and the reactor requires enriched fuel.

7.3.7 Sodium-cooled reactor A nuclear reactor that uses liquid sodium as coolant.

7.3.8 Light-water reactor (LWR) A nuclear reactor in which ordinary water, as distinguished from heavy-water, or a steam/water mixture is used as reactor coolant and moderator. The BWR and PWR are examples of light water reactors.

7.3.9 Primary coolant circuit A system for circulating a primary coolant that serves to withdraw heat from a primary heat source, e.g. from a reactor core or blanket.

7.3.10 Secondary coolant circuit A system for circulating a secondary coolant that serves to withdraw heat from the primary coolant circuit.

7.3.11 Reactor pressure vessel A container, designed to withstand a substantial operating pressure, housing the reactor core and reactor coolant.

7.3.12 Reactor core That region of a reactor that contains the fissile material and is designed to accommodate the nuclear fission chain reaction.

7.3.13 Reflector
1. Part of a reactor placed adjacent to the core of the reactor or to another nuclear-chain-reacting medium to scatter some of the escaping neutrons back into the core or medium;
2. A material or a body of material that reflects incident radiation.

7.3.14 Fuel element The smallest, structurally discrete part of a reactor which has fuel as its principal constituent. Rods, pellets and slugs are characteristic forms of fuel element.

7.3.3 Réacteur à eau pressurisée (P.W.R.) Réacteur dans lequel l'eau servant de fluide de refroidissement et de modérateur est maintenue à une pression telle que l'ébullition ne peut se produire. Ce type nécessite de l'uranium enrichi.

7.3.4 Réacteur à refroidissement au gaz Réacteur dont le fluide de refroidissement est un gaz.

7.3.5 Réacteur à eau lourde Réacteur utilisant de l'eau lourde comme modérateur.

7.3.6 Réacteur à haute température Réacteur utilisant des gaz rares comme réfrigérant dont le coeur utilise des matériaux céramiques et qui fonctionne en régime tel que le réfrigérant soit à des températures élevées.

7.3.7 Réacteur refroidi au sodium Réacteur utilisant le sodium liquide comme réfrigérant.

7.3.8 Réacteur à eau légère Réacteur utilisant de l'eau ou un mélange vapeur-eau comme fluide de refroidissement modérateur.

7.3.9 Circuit primaire de refroidissement Système de circulation du fluide primaire de refroidissement servant à extraire la chaleur d'une source thermique primaire telle qu'un coeur de réacteur ou une couche fertile.

7.3.10 Circuit secondaire de refroidissement, circuit secondaire Système de circulation du fluide secondaire de refroidissement servant à extraire la chaleur du circuit primaire de refroidissement.

7.3.11 Caisson de réacteur, cuve de réacteur Récipient principal entourant le coeur du réacteur avec son réfrigérant.

7.3.12 Coeur Région d'un réacteur contenant la matière fissile et dans laquelle peut se dérouler la réaction en chaîne de fission nucléaire.

7.3.13 Réflecteur
1. Partie du réacteur placée en bordure du coeur ou d'un autre assemblage multiplicateur, en vue de lui renvoyer les neutrons qui s'en échappent.
2. Matériau ou objet qui réfléchit un rayonnement incident de neutrons.

7.3.14 Elément combustible Le plus petit élément ayant une structure propre dans un réacteur nucléaire contenant du combustible nucléaire destiné à être brûlé dans un réacteur. Les éléments combustibles se présentent notamment sous forme de barres, de plaques ou de boulets.

7.3.3 Druckwasserreaktor Ein Druckwasserreaktor ist ein Reaktor, bei dem das als Reaktorkühlmittel dienende Wasser unter so hohem Druck steht, dass die erzeugte Wärme ohne Nettodampfgehalt des Kühlmittels aus dem Reaktor abgeführt wird. Auch dieser Reaktor erfordert angereichertes Uran.

7.3.4 Gasgekühlter Reaktor Ein gasgekühlter Reaktor ist ein mit einem Gas als Reaktorkühlmittel betriebener Reaktor.

7.3.5 Schwerwasserreaktor Ein Schwerwasserreaktor ist ein mit schwerem Wasser als Moderator betriebener Reaktor.

7.3.6 Hochtemperaturreaktor Ein Hochtemperaturreaktor ist ein Reaktor, der unter Verwendung von Edelgasen als Reaktorkühlmittel und weitgehender Verwendung keramischer Werkstoffe im Reaktorkern bei hohen Austrittstemperaturen des Reaktorkühlmittels betrieben wird.

7.3.7 Natriumgekühlter Reaktor Ein natriumgekühlter Reaktor ist ein mit Natrium als Reaktorkühlmittel betriebener Reaktor.

7.3.8 Leichtwasserreaktor Ein Leichtwasserreaktor ist ein mit Wasser oder einem Dampf-Wasser-Gemisch als Reaktorkühlmittel und Moderator betriebener Reaktor.

7.3.9 Primärkühlkreis; Primärkühlkreislauf Der Primärkühlkreis ist das Umlaufsystem für Kühlmittel, das dazu dient, Wärme von einer primären Wärmequelle abzuführen, z.B. aus einer Reaktorspaltzone oder einer Brutzone.

7.3.10 Sekundärkühlkreis; Sekundärkreislauf Der Sekundärkühlkreis ist das Umlaufsystem für Kühlmittel, das dazu dient, Wärme von dem Primärkühlkreis abzuführen.

7.3.11 Reaktordruckbehälter; Reaktordruckgefäss Der Reaktordruckbehälter ist der Behälter, der den Reaktorkern mit Reaktorkühlmittel einschliesst.

7.3.12 Spaltzone Die Spaltzone ist der Bereich eines Reaktors, der im wesentlichen den Spaltstoff enthält und für den Ablauf der Kettenreaktion von Kernspaltungen vorgesehen ist.

7.3.13 Reflektor
1. Der Reflektor ist die Zone in der Nachbarschaft der Spaltzone eines Reaktors oder einer anderen multiplizierenden Anordnung, in der dort ausfliessende Neutronen durch Streuung reflektiert werden.
2. Der Reflektor ist das Material zum Aufbau einer neutronenreflektierenden Zone.

7.3.14 Brennelement Ein Brennelement ist das kleinste, konstruktiv selbständige Bauteil, das Kernbrennstoff zur Verwendung in einem Reaktor enthält. Besondere Formen des Brennelementes sind z.B. Stäbe, Platten, Kugeln.

7.3.3 Reactor de agua a presion (P.W.R.) Reactor donde el agua que sirve de refrigerante y moderador está mantenida a una presión alta para evitar su ebullición manteniéndose líquida. Este tipo requiere combustibles enriquecidos.

7.3.4 Reactor refrigerado por gas Reactor en el que se usa un gas como refrigerante.

7.3.5 Reactor de agua pesada Reactor nuclear que usa agua pesada como moderador.

7.3.6 Reactor de alta temperatura Reactor que con el uso de gases nobles como refrigerante y materiales cerámicos en el núcleo del reactor, funciona a un régimen tal, que el refrigerante se encuentra a temperaturas elevadas.

7.3.7 Reactor refrigerado por sodio Reactor que utiliza sodio líquido como refrigerante.

7.3.8 Reactor de agua ligera Reactor en el que se utiliza, como refrigerante y moderador, agua o una mezcla de agua-vapor.

7.3.9 Circuito primario de refrigeración Sistema de circulación del fluido refrigerante utilizado para extraer el calor primario, como puede ser el del núcleo de un reactor o el de una capa fértil reproductora.

7.3.10 Circuito secundario de refrigeración, circuito secundario Sistema de circulación de fluido refrigerante utilizado para extraer el calor del circuito primario de refrigeración.

7.3.11 Visija de presión de reactor, cuba de reactor Recipiente principal que contiene el núcleo del reactor con su refrigerante.

7.3.12 Núcleo Región del reactor que contiene el material fisible y en la que puede producirse una reacción en cadena de fisión nuclear.

7.3.13 Reflector
1. Parte del reactor situado junto al núcleo o junto a otro conjunto multiplicador con el fin de devolverle los neutrones que tienden a escaparse.
2. Material u objeto que refleja una radiación incidente de neutrones.

7.3.14 Elemento combustible El menor elemento, con estructura propia, en un reactor nuclear que contiene combustible nuclear destinado a ser quemado en un reactor. Los elementos combustibles se presentan principalmente en forma de barras, placas o bolas.

7

7.3.15 Fuel assembly A grouping of fuel elements which is not taken apart during the charging and discharging of a reactor core.

7.3.16 Emergency cooling system A system that in the event of failure of the normal reactor cooling system, e.g. loss of primary coolant, removes the after-heat from the reactor.

7.3.17 Fuel cooling installation; cooling pond (UK) A large container or cell filled with a cooling medium, e.g. water or sodium, in which spent irradiated material, particularly spent nuclear fuel from reactors, is set aside until its activity has decreased to a desired level.

7.3.18 Dousing system; containment spray system A system that reduces the fission product concentration in the reactor containment and thus contributes to lowering temperature and pressure in the building in the event of severe coolant losses.

7.3.19 Core flooding system An emergency cooling system that in the event of failure of the normal reactor cooling system, e.g. loss of primary coolant, removes the after-heat by flooding the reactor core.

7.3.20 Core spray system An emergency cooling system that in the event of failure of the normal reactor cooling system, e.g. loss of primary coolant, removes the after-heat by spraying the reactor core.

7.3.21 Fuel charging machine; refuelling machine Equipment for placing in the reactor core or removing from the reactor core fuel assemblies and other core components, and for their associated transport and handling.

7.3.22 Boric acid system A system for preparing, feeding and recovering boric acid; it serves for controlled modification of the boric acid concentration in the reactor coolant.

7.3.23 Reactor containment A pressure resistant containing system entirely surrounding a nuclear reactor and designed to prevent the release, even under the conditions of a reactor accident, of unacceptable quantities of radioactive material beyond a controlled zone.

7.3.24 Moderator A material used to reduce neutron energy by scattering without appreciable capture.

7.3.25 Reactor coolant A liquid or gas which is circulated through or about the core or blanket of a reactor to remove heat. (See also Primary coolant).

7.3.26 Primary coolant A coolant used to remove heat from a primary source, such as a reactor core or breeding blanket.

7.3.27 Secondary coolant A coolant used to remove heat from the primary coolant circuit.

7.3.28 Reactor protection; reactor protection system; reactor safety system A system that receives information from the various instruments that check the levels of the operating variables essential to reactor security and is able to set in motion one or more safety measures automatically, so as to keep the operation of the reactor within certain limits.

7.3.15 Assemblage combustible Groupement d'éléments combustibles qui restent solidaires au cours du chargement ou du déchargement du coeur d'un réacteur nucléaire.

7.3.16 Système de refroidissement de secours Système assurant l'évacuation de la chaleur résiduelle du coeur du réacteur en cas de défaillance du système normal de refroidissement, par exemple à la suite d'une perte de fluide primaire de refroidissement.

7.3.17 Piscine de désactivation Grand réservoir, ou cellule, généralement rempli d'eau (ou de sodium), dans lequel est entreposé le combustible nucléaire usé jusqu'à ce que son activité ait décrû jusqu'à un niveau souhaité.

7.3.18 Système d'aspersion du bâtiment Système destiné à réduire la teneur en produits de fission dans l'enceinte de sécurité, en cas de perte importante de réfrigérant, contribuant ainsi à abaisser la pression et la température dans le bâtiment.

7.3.19 Noyage du coeur Système de refroidissement de secours qui, en cas de panne du dispositif de refroidissement normal du réacteur (par exemple, en cas de perte de fluide primaire de refroidissement), assure l'évacuation de la chaleur résiduelle par noyage du coeur du réacteur.

7.3.20 Pulvérisation d'eau à coeur Système de refroidissement de secours assurant l'évacuation de la chaleur résiduelle par pulvérisation d'eau dans le coeur du réacteur en cas de défaillance du dispositif de refroidissement normal (par exemple perte de fluide primaire de refroidissement).

7.3.21 Machine de chargement Dispositif destiné à mettre en place les assemblages de combustibles et d'autres composants dans le coeur du réacteur, ou à les retirer, et pouvant assurer au besoin le transport de ces objets.

7.3.22 Dispositif d'injection d'acide borique Dispositif de distribution, d'injection et de récupération de l'acide borique, aux fins de réglage du réacteur.

7.3.23 Enceinte de sécurité Enceinte résistant à la pression entourant un réacteur nucléaire et destinée à empêcher ou à limiter à un montant admissible la dispersion de substances radioactives dans l'atmosphère en cas d'accident de réacteur.

7.3.24 Modérateur Matière utilisée pour réduire l'énergie de neutrons par diffusion sans capture appréciable.

7.3.25 Fluide de refroidissement ou réfrigérant du réacteur Fluide de refroidissement que l'on fait circuler dans un réacteur pour extraire la chaleur du coeur ou de la zone fertile (voir aussi "fluide primaire de refroidissement").

7.3.26 Fluide primaire de refroidissement Fluide de refroidissement utilisé pour extraire la chaleur d'une source primaire, tel qu'un coeur de réacteur ou une couche fertile surrégénératrice.

7.3.27 Fluide secondaire de refroidissement Fluide utilisé pour extraire la chaleur du circuit primaire de refroidissement.

7.3.28 Protection du réacteur Système collectant les informations de différents dispositifs de mesure et de surveillance des paramètres de fonctionnement essentiels à la sécurité d'un réacteur nucléaire, et capable de déclencher automatiquement une ou plusieurs mesures de sauvegarde pour maintenir le régime du réacteur nucléaire dans des limites compatibles avec la sécurité.

7.3.15 Brennelementbündel Ein Brennelementbündel ist eine Anordnung von Brennelementen, die beim Laden und Entladen eines Reaktors eine selbständige Einheit bildet.

7.3.16 Notkühlsystem Das Notkühlsystem eines Reaktors ist ein System, das nach Ausfall der normalen Reaktorkühlung, z.B. beim Verlust von Primärkühlmittel, die Abführung der Nachwärme aus dem Reaktorkern übernimmt.

7.3.17 Abklingbecken Das Abklingbecken ist ein mit einem Kühlmittel, z.B. Wasser oder Natrium, gefülltes Becken, in dem bestrahlte Stoffe, insbesondere Kernbrennstoffe aus Reaktoren, so lange lagern, bis ihre Aktivität auf einen gewünschten Wert abgenommen hat.

7.3.18 Gebäudesprühsystem Das Gebäudesprühsystem eines Kernkraftwerkes ist ein System, das nach grösseren Kühlmittelverlusten die Spaltproduktkonzentration in der Atmosphäre der Sicherheitshülle verringert und dazu beiträgt, Druck und Temperatur im Gebäude abzusenken.

7.3.19 Kernflutsystem Das Kernflutsystem ist ein Notkühlsystem, das nach Ausfall der normalen Reaktorkühlung, z.B. bei Verlust von Primärkühlmittel, durch Fluten des Reaktorkerns die Abführung der Nachwärme übernimmt.

7.3.20 Kernsprühsystem Der Kernsprühsystem ist ein Notkühlsystem, das nach Ausfall der normalen Reaktorkühlung, z.B. bei Verlust von Primärkühlmittel, durch Besprühen des Reaktorkers die Abführung der Nachwärme übernimmt.

7.3.21 Lademaschine Eine Lademaschine ist eine Vorrichtung zum Einsetzen und Herausnehmen von Brennelementbündeln und anderen Kernbauteilen in den oder aus dem Kern eines Reaktors, gegebenenfalls auch zum Transport dieser Gegenstände zwischen Reaktor und Lagerbecken bzw, Schleuse und zu ihrer Handhabung im Lagerbecken.

7.3.22 Borsäuresystem Das Borsäuresystem eines Reaktors ist ein System zur Bereitstellung, Einspeisung und Rückgewinnung von Borsäure; es dient zur gesteuerten Aenderung der Borsäurekonzentration im Kühlmittel einer Reaktoranlage.

7.3.23 Sicherheitshülle; Sicherheitsbehälter Die Sicherheitshülle ist die druckfeste Umhüllung der Reaktoranlage, die das Austreten unzulässiger Mengen radioaktiver Stoffe in die freie Umgebung vor allem bei einem Reaktorstörfall verhindert.

7.3.24 Moderator Ein Moderator ist ein zur Moderation von Neutronen geeigneter Stoff.

7.3.25 Reaktorkühlmittel Das Reaktorkühlmittel ist ein Medium, das zum Abführen der Wärme aus der Reaktorspaltzone oder Brutzone dient. (siehe auch Primärkühlmittel).

7.3.26 Primärkülmittel Das Primärkühlmittel ist ein Medium, das zum Abführen der Wärme aus einer Primärquelle, z.B. der Reaktorspaltzone oder der Brutzone, dient.

7.3.27 Sedundärkühlmittel Das Sekundärkühlmittel ist ein Medium das zum Abführen der Wärme des Primärkühlkreises dient.

7.3.28 Reaktorschutz Als Reaktorschutz bezeichnet man ein System, das Informationen von verschiedenen Messeinrichtungen erhält, die die für die Sicherheit wesentlichen Betriebsgrössen eines Kernreaktors überwachen, und das imstande ist, automatisch eine oder mehrere Sicherheitsmassnahmen auszulösen, um den Zustand des Kernreaktors in sicheren Grenzen zu halten.

7.3.15 Conjunto combustible Grupo de elementos combustibles que permanecen solidarios en la carga o descarga del núcleo del reactor nuclear.

7.3.16 Sistema de refrigeración de emergencia Sistema que, en el caso de avería en el sistema de refrigeración normal del reactor, por ejemplo, pérdida del refrigerente primario, asegura la evacuación del calor residual del núcleo.

7.3.17 Piscina de desactivación Gran recipiente o célula, generalmente llena de agua, (o de sodio) en que se deposita el combustible nuclear utilizado, hasta que su radiactividad disminuya hasta un nivel deseado.

7.3.18 Sistema de aspersión Sistema destinado a reducir la concentración en productos de fisión en el recinto de seguridad, en caso de pérdidas importantes de refrigerante, contribuyendo así a reducir la presión y la temperatura en el edificio.

7.3.19 Sistema de inundación del núcleo Sistema de refrigeración de emergencia que, en el caso de avería en el sistema de refrigeración normal del reactor, (por ejemplo, pérdida del refrigerante primario) asegura la evacuación del calor mediante la inundación del núcleo.

7.3.20 Sistema de rociado del núcleo Sistema de refrigeración de emergencia que, en el caso de avería en el sistema de refrigeración normal del reactor, (por ejemplo, pérdida del refrigerador primario), absorbe el calor por medio de rociado del núcleo.

7.3.21 Máquina de carga Equipo destinado a introducir en el núcleo del reactor o quitar del núcleo del reactor los conjuntos de combustibles y de otros componentes y que pueden asegurar su transporte.

7.3.22 Sistema de ácido bórico Sistema para preparar, alimentar y recuperar ácido bórico; sirve para la modificación controlada de la concentración del ácido bórico en el refrigerante del reactor.

7.3.23 Recinto de seguridad Recinto que encierra por completo el reactor nuclear, que resiste la presión y que está destinado a impedir o a limitar a dosis admisibles la dispersión de sustancias reactivas en la atmósfera, en caso de accidente del reactor.

7.3.24 Moderador Material utilizado para reducir la energía cinética de los neutrones, por difusión y sin captura apreciable.

7.3.25 Fluido de refrigeración o refrigerante del reactor Liquido o gas que circula por, o al rededor del núcleo de un reactor para extraer su calor o el de la zona fértil (vease también "refrigerante primario").

7.3.26 Refrigerante primario Refrigerante usado para extraer calor de una fuente primaria, tal como el núcleo de un reactor o de una capa fértil reproductora.

7.3.27 Refrigerante secundario Refrigerante usado para extraer calor del circuito primario de refrigeración.

7.3.28 Sistema de seguridad Sistema que recibe información de varios instrumentos de medida y alarma, de las variables en funcionamiento, esenciales para la seguridad del reactor y puede poner en marcha, automáticamente, una o más medidas de seguridad, para mantener el funcionamiento del reactor nuclear dentro de límites compatibles con la seguridad.

7

7.3.29 Air discharge purification system; air filtration system Equipment for removing radioactive impurities from the air in the controlled area of nuclear plants.

7.3.30 Shield Protective material intended to reduce the intensity of radiation entering a region.

7.3.31 Breeder element The smallest structurally discrete part of a reactor which has fertile material as its principal constituent.

7.3.32 Breeder assembly; breeder element assembly A grouping of breeder elements which is not taken apart during the charging and discharging of a reactor core.

7.3.33 Blanket A region of fertile material placed around or within a reactor core for the purpose of conversion. By extension the term ''blanket'' may be used when the purpose is transformation of non-fertile material.

7.3.34 Burnable poison A neutron absorber (or poison) purposely included in a reactor which by its progressive burnup helps to compensate for loss of reactivity as the nuclear fuel in the reactor is consumed.

7.3.35 Reactor control system Equipment for varying the reaction rate in a reactor or adjusting reactivity to achieve or maintain a desired state of operation.

7.3.36 Neutron absorber An object, incorporated as a component of a reactor, with which neutrons interact significantly or predominantly by reactions resulting in their disappearance as free particles without production of other neutrons.

7.3.37 Control member; control element A movable part of a reactor which itself affects reactivity and is used for reactor control. A ''control rod'' is a control member in the form of a rod.

7.3.38 Safety member; safety element A control member which singly or in concert with others provides a reserve of negative reactivity for the purpose of emergency shutdown of a reactor. A ''safety rod'' is a safety member in which the control member is in the form of a rod.

7.4 Operating Parameters of Power Reactors

7.4.1 Burnup Induced nuclear transformation of atoms during reactor operation. The term may be applied to fuel or other materials.

7.4.2 Specific burnup; fuel irradiation level The quotient of the total energy released by a nuclear fuel and the initial mass of the nuclear fuel. It is commonly expressed in megawatt-days per tonne.

7.4.3 Shutdown reactivity The reactivity of the reactor when it has been reduced to the sub-critical state by normal operating procedures; shutdown reactivity is always negative.

7.4.4 Reactivity worth The change in reactivity brought about by altering the position of a reactor component, or of an object or material introduced into a reactor, or by changing an operating variable.

7.3.29 Système d'épuration de l'air de rejet Dispositif qui, dans un réacteur, élimine les impuretés radioactives de l'air de la zone contrôlée.

7.3.30 Bouclier Matériau de protection destiné à réduire l'intensité du rayonnement ionisant pénétrant dans une région.

7.3.31 Elément fertile Le plus petit élément formant un tout, contenant la matière fertile destinée à être brulée dans un réacteur.

7.3.32 Assemblage fertile Groupement d'éléments fertiles qui restent solidaires au cours du chargement ou du déchargement d'un réacteur.

7.3.33 Couche fertile Région de matière fertile placée autour ou à l'intérieur du coeur d'un réacteur pour assurer la conversion de cette matière.

7.3.34 Poison consommable Absorbant neutronique introduit à dessein dans un réacteur, pour contribuer au contrôle des variations à long terme de la réactivité, au moyen de sa combustion progressive.

7.3.35 Ensemble de règlage automatique (pilote automatique) Ensemble des dispositifs permettant d'atteindre ou de maintenir un état de régime donné.

7.3.36 Absorbant neutronique, absorbant de neutrons Objet dont l'interaction avec les neutrons donne lieu à des réactions provoquant leur disparition en tant que particules libres sans production d'autres neutrons.

7.3.37 Elément de commande Partie mobile d'un réacteur dont l'action influe sur la réactivité et qui est utilisée en vue de la commande du réacteur.

7.3.38 Elément de sécurité Élément de règlage qui, seul ou de concert avec d'autres, fournit une réserve de réactivité négative pour le cas d'un arrêt d'urgence d'un réacteur.

7.4 Comportement en Service des Réacteurs de Puissance

7.4.1 Combustion nucléaire Transformation nucléaire d'atomes induite pendant le fonctionnement d'un réacteur thermique.
Remarque : ce terme peut être appliqué au combustible ou à d'autres matières.

74.2 Combustion massique Énergie totale libérée par unité de masse d'un combustible nucléaire. C'est le quotient de l'énergie nucléaire par la masse initiale du combustible et s'exprime, communément, en mégawatt-jours par tonne.

7.4.3 Réactivité résiduelle Réactivité d'un réacteur amené à un régime sous-critique par des moyens normaux. La réactivité résiduelle est toujours négative.

7.4.4 Equivalent de réactivité Variation de réactivité résultant du changement de position d'un élément du réacteur ou de celle d'un objet ou d'un matériau introduit dans le réacteur ou de la modification d'un paramètre d'exploitation.

7.3.29 Abluftreinigungssystem Das Abluftreinigungssystem einer kerntechnischen Anlage ist ein System zum Abscheiden radioaktiver Verunreinigungen aus der Luft des Kontrollbereichs dieser Anlagen.

7.3.30 Abschirmung Eine Abschirmung ist eine Anordnung von Stoffen, die dem Zweck dient, die in einen bestimmten Bereich gelangende ionisierende Strahlung zu verringern.

7.3.31 Brutelement Ein Brutelement ist das kleinste konstruktiv selbständige Bauteil, das Brutstoff zur Verwendung in einem Reaktor enthält.

7.3.32 Brutelementbündel Ein Brutelementbündel ist eine Anordnung aus Brutelementen, die beim Laden und Entladen eines Reaktors eine selbständige Einheit bildet.

7.3.33 Brutzone Die Brutzone ist eine Reaktorzone ausser- oder innerhalb der Spaltzone, die Brutstoffe zum Zweck der Konversion enthält.

7.3.34 Reaktorgift, abbrennbares Ein abbrennbares Reaktorgift ist ein in einem Reaktor eingebrachter Neutronenabsorber, der durch seine allmähliche Aufzehrung den Reaktivitätsverlust durch den Abbrand des Kernbrennstoffs ausgleichen soll.

7.3.35 Reaktorsteuer- und -regelsystem Als Reaktorsteuer- und regelsystem bezeichnet man die Gesamtheit der Einrichtungen zum Erreichen oder Einhalten eines gewünschten Betriebszustandes.

7.3.36 Absorberelement Ein Absorberelement ist ein Neutronenabsorber enthaltendes Bauteil eines Reaktors, das der Beeinflussung der Reaktivität oder der Reaktivitätsverteilung dient, ohne Erzeugung anderer Neutronen.

7.3.37 Stellmittel Ein Stellmittel ist ein Funktionselement eines Kernreaktors, der direkt die Reaktivität beeinflusst und der Reaktorsteuerung oder -regelung dient.

7.3.38 Sicherheitselement Das Sicherheitselement ist ein Stellmittel, das allein oder zusammen mit anderen eine negative Reaktivitätsreserve für die Notabschaltung eines Kernreaktors liefert.

7.4 Betriebsverfahren von Leistungsreaktoren

7.4.1 Abbrand Der Abbrand ist die infolge des Reaktorbetriebes bewirkte Umwandlung von Atomkernen.
Anmerkung : Der Begriff kann sich auf Kernbrennstoff oder andere Stoffe beziehen.

7.4.2 Abbrand, spezifischer Als spezifischen Abbrand bezeichnet man den Quotienten aus der in einem Kernbrennstoff durch Abbrand freigesetzten Gesamtenergie und der ursprünglich vorhandenen Kernbrennstoffmasse.

7.4.3 Abschaltreaktivität Die Abschaltreaktivität ist die Reaktivität des durch Abschalten mit betriebsüblichen Mitteln in den unterkritischen Zustand gebrachten Reaktors. Die Abschaltreaktivität ist stets negativ.

7.4.4 Reaktivitätsäquivalent ist die durch die Aenderung der Lage eines Reaktorteils oder eines in den Reaktor eingebrachten Gegenstandes bzw. Materials oder durch die Aenderung einer Betriebsgrösse verursachte Reaktivitätsänderung.

7.3.29 Sistema de descontaminación del aire Equipo empleado para extraer impurezas radiactivas del aire en la zona controlada de instalaciones nucleares.

7.3.30 Blindaje Conjunto de materiales destinados a reducir la intensidad de las radiaciones ionizantes que penetran en una región.

7.3.31 Elemento fértil Componente independiente, el más pequeño estructuralmente, del material fértil usado en un reactor.

7.3.32 Haz de elementos fértiles Agrupación de elementos fértiles que permanecen juntos durante la carga y descarga del núcleo de un reactor.

7.3.33 Capa fértil Región de material fértil, colocado alrededor o en el interior del núcleo de un reactor, para asegurar la conversión de este material.

7.3.34 Veneno consumible Absorbente neutrónico introducido intencionadamente en el reactor para contribuir al control de las variaciones, a largo plazo, de la reactividad, debido a su destrucción progresiva.

7.3.35 Conjunto de regulacion automática Conjunto de dispositivos que permiten alcanzar o mantener un régimen de funcionamiento dado.

7.3.36 Absorbente de neutrones Sustancia cuya interacción con los neutrones da lugar a reacciones que provocan la desaparición de neutrones, sin producción de otros neutrones.

7.3.37 Elemento de control Parte móvil de un reactor que por sí misma afecta a la reactividad y que se utiliza para el control del reactor.

7.3.38 Elemento de seguridad Elemento de control que, solo o en combinación con otros, proporciona una reserva de reactividad negativa para producir la parada de emergencia de un reactor.

7.4 Comportamiento en Servicio de los Reactores de Potencia

7.4.1 Combustión nuclear Transformación nuclear de átomos, producida durante el funcionamiento de un reactor. Puede aplicarse este término al combustible o a otros materiales.

7.4.2 Grado de quemado específico Energía total liberada por unidad de masa de un combustible nuclear. Es el cociente entre la energía total liberada en un combustible nuclear y la masa inicial del combustible. Se expresa normalmente en megavatios-dias por tonelada.

7.4.3 Reactividad residual Reactividad de un reactor cuando ha sido reducida al estado de subcrítioo por procesos de operación normales. La reactividad residual es siempre negativa.

7.4.4 Reactividad equivalente Variación de reactividad ocasionada por alterar la posición de un elemento de un reactor, o de un objeto o de un material introducido en el reactor, o por la modificación de una variable de explotación.

7

7.4.5 Reactivity balance The balance between excess reactivity referred to a specific reference state of a reactor and the values of reactivity worth that result from the change of reactor state with reference to the reference state.

Note. As "reference state" may be chosen the state of the cold reactor with a specific core at the commencement of initial start-up (preferably in cases where safety considerations are concerned), but also any other operating state.

7.4.6 Excess reactivity The maximum reactivity attainable at any time by adjustment of the control members.

7.4.7 Reactivity coefficient The partial derivative of reactivity with respect to some specified parameter that influences reactivity (e.g. temperature or pressure).

7.4.8 Reactor time constant; reactor period The time, T, required for the neutron flux density, ϕ, in a reactor to change by a factor e = 2.718. . ., when the flux density is rising or falling exponentially. Generally, however, T is defined as:

$$\frac{1}{T} = \frac{1}{\phi} \cdot \frac{d\phi}{dt}$$

7.4.9 Power density The power generated per unit volume of a reactor core.

7.4.10 Xenon poisoning effect; xenon effect The reduction in reactivity caused by neutron capture in Xenon-135, a fission product which is a nuclear poison.

7.4.11 Fuel rating The quotient of the total thermal power evolved in a reactor core and the initial mass of fissile and fertile nuclides. Sometimes the quotient is formed with the mass of the initial charge.

7.4.12 Residual heat; after-heat (1) For a shutdown reactor, the heat resulting from residual radioactivity and fission. (2) For reactor fuel or reactor components after removal from the reactor, the heat resulting from residual radioactivity.

7.4.13 Linear power density The thermal power generated in a fuel element divided by the length of the element.

7.4.14 Design basis accident An accident in an installation that, by agreement, needs to be taken into account in devising protective measures at the design stage.

7.4.15 Maximum credible accident The worst accident in a reactor or nuclear energy installation that, by agreement, need be taken into account in devising protective measures at the design stage.

7.4.16 Critical heat flux (DNB heat flux) The local heat flux density between a surface and a cooling liquid which gives a maximum in the curve of heat flux density against temperature difference, associated with the change from nucleate boiling to film boiling. (Also called DNB (Departure from Nucleate Boiling) heat flux.)

7.4.5 Bilan de réactivité Comparaison entre l'excédent de réactivité relatif à un état de référence déterminé du réacteur et les équivalents de réactivité résultant d'une modification de l'état de référence.

Remarque : l'état de référence choisi peut être celui du réacteur froid avec un coeur défini au début de la première mise en service (de préférence s'il s'agit de considérations de sécurité), ou tout autre état de régime.

7.4.6 Excédent de réactivité (réactivité excédentaire) Réactivité maximale disponible à tout moment par ajustement des éléments de commande.

7.4.7 Coefficient de Réactivité Dérivée partielle de la réactivité par rapport à un paramètre donné qui l'influence (température ou pression, par exemple).

7.4.8 Constante de temps d'un réacteur - période d'un réacteur Temps T nécessaire pour que la densité de flux de neutrons ϕ dans un réacteur varie d'un facteur e = 2,718 . . . , lorsque la densité de flux augmente ou diminue de façon exponentielle. T est généralement défini par :

$$\frac{1}{T} = \frac{1}{\phi} \cdot \frac{d\phi}{dt}$$

7.4.9 Puissance volumique (du réacteur) Puissance engendrée par unité de volume du coeur.

7.4.10 Empoisonnement xénon (effet xénon) Reduction de la réactivité provoquée par la capture de neutrons par le produit de fission Xe - 135, qui est un poison nucléaire.

7.4.11 Puissance spécifique du combustible Quotient de la puissance thermique totale développée dans le coeur d'un réacteur par la masse initiale des nucléides fissiles et fertiles. En général, le quotient se rapporte à la masse de la charge initiale.

7.4.12 Chaleur résiduelle
1. Pour un réacteur à l'arrêt, chaleur résultant de la radio-activité et des fissions résiduelles.
2. Pour le combustible ou les composants extraits d'un réacteur, chaleur résultant de la radio-activité résiduelle.

7.4.13 Puissance (linéaire) du barreau Puissance dégagée par un élément de barreau de combustible divisée par la longueur de cet élément.

7.4.14 Incident technique théorique Incident qui, par convention, doit être pris en considération dès la conception d'une installation nucléaire pour établir des mesures permettant de s'assurer le contrôle de la situation.

7.4.15 Accident maximal prévisible Accident le plus grave d'un réacteur ou d'une installation nucléaire qui, par convention, doit être pris en considération pour établir les mesures de protection.

7.4.16 Flux de caléfaction Valeur locale de la densité de flux calorifique entre une surface et un liquide de refroidissement qui donne, dans la courbe de la densité de flux calorifique en fonction de la différence de température, un maximum correspondant au changement de l'ébullition par bulle en ébullition par film.

Remarque : on parle également de flux de caléfaction DNB (Départure from Nucleate Boiling).

7.4.5 Reaktivitätsbilanz Die Reaktivitätsbilanz ist die Gegenüberstellung der auf einen bestimmten Referenzzustand eines Reaktors bezogenen Ueberschussreaktivität und der Reaktivitätsäquivalente, die sich aus der Aenderung des Reaktorzustandes in bezug auf den Referenzzustand ergeben.

Anmerkung : Als Referenzzustand kann der Zustand des kalten Reaktors mit einer bestimmten Spaltzone zu Beginn der ersten Inbetriebnahme (vorzugsweise bei Sicherheitsbetrachtungen), aber auch jeder andere Betriebszustand gewählt werden.

7.4.6 Ueberschussreaktivität Die Ueberschussreaktivität ist die höchste verfügbare Reaktivität, die mit betriebsüblichen Mitteln erreichbar ist.

7.4.7 Reaktivitätskoeffizient Der Reaktivitätskoeffizient ist der Differentialquotient der Reaktivität nach einer sie beeinflussenden Grösse (z.B. Temperatur, Druck).

7.4.8 Reaktorzeitkonstante; Reaktorperiode Die Reaktorzeitkonstante ist die Zeit T, in der die Neutronenflussdichte ϕ in einem Reaktor sich um den Faktor e = 2,718. . . ändert, wenn sie exponentiell zu oder abnimmt. Im allgemeinen ist jedoch T definiert durch

$$\frac{1}{T} = \frac{1}{\phi} \cdot \frac{d\phi}{dt}$$

7.4.9 Leistungsdichte, mittlere (des Reaktors) Die mittlere Leistungsdichte ist der Quotient aus der thermischen Leistung und dem Volumen der Spaltzone.

7.4.10 Xenonvergiftung Unter Xenonvergiftung versteht man die Verminderung der Reaktivität verursacht durch Neutroneneinfang im Spaltprodukt Xe-135.

7.4.11 Brennstoffleistung, spezifische Die spezifische Brennstoffleistung ist die in der Spaltzone eines Reaktors in einem gegebenen Volumenelement des Kernbrennstoffs erzeugte thermische Leistung, dividiert durch die in diesem Volumenelement enthaltene Masse an Kernbrennstoff. Bezieht sich in der Regel auf der Erstkern.

7.4.12 Nachwärme Als Nachwärme bezeichnet man
1. bei einem abgeschalteten Reaktor die Wärme, die von der restlichen Radioaktivität oder Spaltung herrührt.
2. die Wärme, die aus der restlichen Radioaktivität im Brennstoff oder in Komponenten entsteht, nachdem diese aus dem Reaktor entfernt worden sind.

7.4.13 Stableistung, lineare Die lineare Stableistung ist die in einem Langenelement eines Brennstabes freigesetzte Leistung, dividiert durch dieses Längenelement.

7.4.14 Auslegungsstörfall Der Auslegungsstörfall ist ein Störfall, für den gemäss Uebereinkunft bei der Auslegung der Anlage Massnahmen getroffen werden müssen, die seine Beherrschung sicherstellen.

7.4.15 Grösster anzunehmender Unfall (GAU) Der grösste anzunehmende Unfall (GAU) ist der schwerste Störfall in einer kerntechnischen Anlage, für den gemäss Uebereinkunft bei der Auslegung der Anlage Massnahmen getroffen werden müssen, die seine Beherrschung sicherstellen.

7.4.16 Wärmestromdichte, kritische Die kritische Wärmestromdichte ist die örtliche Wärmestromdichte zwischen einer Oberfläche und einer Kühlflüssigkeit im Maximum der Kurve der Wärmestromdichte über der Temperaturdifferenz. Damit ist ein Wechsel vom Keimsieden zum Filmsieden verbunden.

Anmerkung : Auch DNB-Wärmestromdichte (Departure from Nucleare Boiling) genannt.

7.4.5 Balance de radiactividad Comparación entre el excedente de reactividad relativo, en un estado de referencia determinado del reactor y los equivalentes de reactividad resultantes de una modificación del estado del reactor con relación al estado de referencia.

Observación: El estado de referencia elegido puede ser el del reactor frío con un núcleo definido al comienzo de la primera puesta en servicio (preferentemente si se trata de consideraciones de seguridad) o cualquier otro régimen de funcionamiento.

7.4.6 Exceso de reactividad Reactividad máxima obtenible·en cualquier momento mediante el ajuste de los elementos de control.

7.4.7 Coeficiente de reactividad Derivada parcial de la reactividad con relación a un parámetro específico, que influye sobre ella (por ejemplo, con relación a la temperatura, la presión etc.).

7.4.8 Constante de tiempo de un reactor; periodo de un reactor Tiempo T necesario para que la densidad de flujo de neutrones en un reactor, varie en el factor e = 2.718. . . . cuando el flujo aumenta o disminuye de forma exponencial, T queda definido, generalmente, por:

$$\frac{1}{T} = \frac{1}{\phi} \cdot \frac{d\phi}{dt}$$

7.4.9 Densidad de potencia (del reactor) Potencia producida, por unidad de volumen del núcleo del reactor.

7.4.10 Envenenamiento por el xenon (efecto xenon) Reducción de la radiactividad provocada por la captura de neutrones por el producto de fusión Xe -135, que es un veneno nuclear.

7.4.11 Potencia específica del combustible Cociente entre la potencia térmica total desarrollada en el núcleo de un reactor y la masa inicial de los nucleıdos fisibles y fértiles. A veces el cociente se forma con la masa de la carga inicial.

7.4.12 Calor residual
1. Referido a un reactor tras su parada: calor resultante de la radiactividad y de las fisiones residuales.
2. Referido al combustible o a los componentes extraídos del reactor: calor resultante de la radioactividad residual.

7.4.13 Potencia lineal de una barra Es la potencia térmica engendrada en una barra combustible dividido por la longitud de dicho elemento.

7.4.14 Incidente técnico teórico Posible incidente hipotético en la instalación que, convencionalmente, se necesita considerar cuando se proyectan medidas que permitan asegurar el control de la situación.

7.4.15 Máximo accidente previsible El accidente más grave en un reactor o en una instalación nuclear, que se necesita tomar en cuenta cuando se proyectan convencionalmente medidas de seguridad.

7.4.16 Flujo calorífico crítico Valor local de la densidad del flujo calorífico entre una superficie y un liquido refrigerante, que da en la curva de la densidad de flujo calorífico, en función de la diferencia de temperatura, un máximo correspondiente a un cambio de la ebullición de burbuja en ebullición por película.

7

7.4.17 Reactor thermal power The heat generated in a nuclear reactor in unit time.

7.4.18 Emergency shutdown; scram The act of shutting down a reactor suddenly to prevent or minimize a dangerous condition.

7.5 Radiation Protection and Environmental Effects

7.5.1 Radiation protection All measures associated with the limitation of the harmful effects of ionizing radiation on people and all measures designed to limit radiation-induced chemical and physical damage to materials.

7.5.2 Dose A general term denoting the quantity of radiation or energy absorbed. For special purposes, it must be appropriately qualified.
Note. The term ''dose'' has been used with a variety of specific meanings, such as absorbed dose, exposure and fluence, but such uses are to be avoided.

7.5.3 Dose rate The ratio of the dose in a suitably small time interval to the time interval.

7.5.4 Absorbed dose The absorbed dose, D, is the quotient of $d\bar{\varepsilon}$ by dm, where $d\bar{\varepsilon}$ is the mean energy imparted by ionizing radiation to the matter in a volume element and dm is the mass of the matter in that volume element.

$$D = \frac{d\bar{\varepsilon}}{dm}$$

The special unit of absorbed dose is the rad.
$$1 \, \text{rad} = 10^{-2} \text{J kg}^{-1}$$

7.5.5 Exposure (1) For X or gamma radiation in air, the sum of the electrical charges of all of the ions of one sign produced in air when all electrons liberated by photons in a suitably small element of volume of air are completely stopped in air, divided by the mass of the air in the volume element. It is commonly expressed in roentgen. (2) The incidence of radiation on living or inanimate material, by accident or intent.
Note. To avoid confusion, meaning (2) of the above term should be avoided wherever possible.

7.5.6 Quality factor A factor depending on the linear energy transfer in water of primary or secondary charged particles, by which absorbed dose is multiplied to obtain, according to practice in the field of radiation protection, an evaluation on a common scale, for all ionizing radiations, of the irradiation incurred by exposed persons.

7.5.7 Dose equivalent The product of absorbed dose, quality factor, distribution factor, and other modifying factors necessary to obtain an evaluation of the effects of irradiation received by exposed persons, so that the different characteristics of the exposure are taken into account. It is commonly expressed in rems.

$$1 \, \text{rem} = 10^{-2} \text{J kg}^{-1}$$

7.4.17 Puissance thermique du réacteur Quantité de chaleur libérée dans le coeur par unité de temps.

7.4.18 Arrêt d'urgence Action d'arrêter brusquement un réacteur pour éviter une situation dangereuse ou en réduire les effets.

7.5 Radioprotection et Impact Écologique

7.5.1 Protection radiologique Création d'installations et exécution des mesures de protection contre les effets nocifs des rayons ionisants.

7.5.2 Dose de rayonnement Terme général désignant une quantité de rayonnement ou d'énergie absorbée (cf, ''dose absorbée'', ''dose ionique'' et ''équivalent de dose'', par exemple).

7.5.3 Débit de dose, taux de dose Quotient de la dose pendant un intervalle de temps suffisamment court par la durée de cet intervalle.

7.5.4 Dose absorbée La dose absorbée D émise par un rayonnement ionisant est égale au quotient de $\varrho \, dW_D$ à dm, dW_D représentant l'énergie communiquée par le rayonnement à un volume dV de la matière et $dm = \varrho dV$ où dm désigne la masse du matériau de densité ϱ contenu dans ce volume

$$D = \frac{dW_D}{dm} = \frac{1}{\varrho} \cdot \frac{dW_D}{dV}$$

Unité : $1 \, \text{rad} = 10^{-2} \text{J/kg}$

7.5.5 Exposition, dose ionique L'exposition provoquée par le rayonnement X ou gamma est égale au quotient de la somme des charges électriques des ions de même signe formés immédiatement ou indirectement par rayonnement dans un volume d'air donné par la masse d'air contenue dans ce volume.

7.5.6 Facteur de qualité Facteur dépendant du transfert linéique d'énergie dans l'eau de particules chargées primaires ou secondaires, par lequel il faut multiplier la dose absorbée pour obtenir, à l'usage de la radioprotection, une évaluation, à une échelle commune à tous les rayonnements ionisants, de l'irradiation reçue par les personnes exposées.

7.5.7 Équivalent de dose L'Équivalent de dose est le produit de la dose absorbée par le facteur de qualité correspondant.
Unité : Joule/kilogramme (J/kg)
Autre unité : rem, $1 \, \text{rem} = \frac{1}{100} \text{J/kg}$

7.4.17 Reaktorleistung, thermische Wärmemenge, die in der Zeiteinheit im Kernreaktor freigesetzt wird.

7.4.18 Schnellabschaltung Die Schnellabschaltung ist das schnelle Abschalten eines Reaktors zur Verhinderung oder Begrenzung eines unzulässigen Betriebszustandes.

7.5. Strahlenschutz und Umweltbeeinflussung

7.5.1 Strahlenschutz Unter Strahlenschutz versteht man die Schaffung von Einrichtungen und die Durchführung von Massnahmen zum Schutz vor schädlichen Wirkungen ionisierender Strahlen.

7.5.2 Strahlendosis; Dosis Die Strahlendosis ist ein Mass für eine näher anzugebende Strahlenwirkung (siehe z.B. Energiedosis, Ionendosis, Aequivalentdosis).

7.5.3 Dosisleistung; Dosisrate Die Dosisleistung ist der Quotient aus der Dosis in einem angemessen kleinen Zeitintervall und diesem Zeitintervall.

7.5.4 Energiedosis Die von einer ionisierenden Strahlung erzeugte Energiedosis D ist der Quotient aus dW_D und dm, wobei dW_D die Energie ist, die auf das Material in einem Volumenelement dV durch die Strahlung übertragen wird, und $dm = \varrho dV$ die Masse des Materials mit der Dichte ϱ in diesem Volumenelement

$$D = \frac{dW_D}{dm} = \frac{1}{\varrho} \cdot \frac{dW_D}{dV}$$

Einheit : 1 Rad = 10^{-2} J/kg

7.5.5 Ionendosis Die durch Röntgen- oder Gammastrahlung hervorgerufene Ionendosis ist der Quotient aus der elektrischen Ladung der Ionen eines Vorzeichens, die in Luft in einem Volumenelement durch Strahlung unmittelbar oder mittelbar gebildet werden, und der Masse der Luft in diesem Volumenelement.

7.5.6 Bewertungsfaktor, Qualitätsfaktor Der Bewertungsfaktor ist das Produkt von festzulegenden Faktoren, mit dem bei der Berechnung der Aequivalentdosis die Energiedosis multipliziert wird, um dem jeweiligen Strahlenrisiko für die verschiedenen ionisierenden Strahlungsarten und den jeweiligen Bestrahlungsbedingungen Rechnung zu tragen.

7.5.7 Aequivalentdosis Die Aequivalentdosis ist das Produkt aus der Energiedosis und dem jeweiligen Bewertungsfaktor.
Einheit : Joule/Kilogramm (J/kg)
andere Einheit : rem; 1 rem = $\frac{1}{100}$ J/kg

7.4.17 Potencia termica del reactor Cantidad de calor producida en un reactor nuclear en una unidad de tiempo.

7.4.18 Parada de emergencia Acción de parada brusca de un reactor para evitar o minimizar una situación peligrosa.

7.5 Protección radiologica y Efectos Ecológicos

7.5.1 Proteccion radiológica Creación de instalaciones y realización de medidas de protección contra los efectos nocivos de las radiaciones ionizantes.

7.5.2 Dosis de radiación Término general que indica la cantidad de radiación o energía absorbida. Para usos particulares debe precisarse (vease p. ej. "dosis absorbida", "dosis iónica", y "dosis equivalente").

7.5.3 Dosis por unidad de tiempo Relación de la dosis, en un intervalo de tiempo apropiadamente pequeño, a dicho intervalo de tiempo.

7.5.4 Dosis absorbida La dosis absorbida D emitida por una radiación ionizante es igual al cociente entre dW_D y dm, siendo dW_D la energía comunicada por una radiación a un volumen dV del material, y $dm = \varrho dV$ siendo dm la masa del material de densidad ϱ contenida en el volumen:

$$D = \frac{dW_D}{dm} = \frac{1}{\varrho} \cdot \frac{dW_D}{dV}$$

Unidad: 1 Rad. = 10^{-2} J/kg.

7.5.5 Exposición, dosis iónica La exposición provocada por la radiación X o γ es igual al cociente entre la suma de las cargas eléctricas de todos los iones del mismo signo, formadas directa o indirectamente por radiación en un volumen dado de aire, y la masa de aire contenida en dicho volumen.

7.5.6 Factor de calidad Factor que depende de la transferencia lineal de energía en el agua de las partículas cargadas primarias o secundarias, por el cual es necesario multiplicar la dosis absorbida, para obtener en el empleo de la protección contra las radiaciones, una evaluación, en una escala común a todas las radiaciones ionizantes, de la radiación recibida por las personas expuestas.

7.5.7 Dosis equivalente Producto de la dosis absorbida por el factor de calidad correspondiente.
Unidad: Julio — Kilogramo (J/kg)
Otra unidad: rem.; 1 rem = $\frac{1}{100}$ J/kg.

7

7.5.8 Maximum permissible dose equivalent (MPDE) The largest dose equivalent received within a specified period which is permitted by a regulatory committee on the assumption that there is no appreciable probability of somatic or genetic injury. Different levels of MPDE may be set for different groups within a population. Also called Maximum permissible dose (MPD).

7.5.9 The incidence of man-made or of both natural and man-made ionizing radiation on persons, groups of the population or the whole population. No single English term or phrase defines this concept exactly, but see term 7.5.5 (2) above.

7.5.10 Individual dose The dose (exposure, absorbed dose or dose equivalent) to the body or to a given critical organ received by any individual during a given period of time.

7.5.11 Group/sub-population collective dose A component of the population dose (see 8.5.8) related to a given sub-population, which, for some purposes, may be the population of a country or region. The group/sub-population collective dose is measured in rems.

7.5.12 Radioactive fall-out The deposition upon the surface of the earth of radioactive substances from the explosion of a nuclear device or from their accidental release.

7.5.13 Radiotoxicity The toxicity attributable to ionizing radiation emitted by an incorporated radionuclide and its daughters; radiotoxicity is related not only to the radioactive characteristics of this radionuclide but also to its chemical and physical state and to metabolism of this element in the body or in the organ.

Note. According to their relative radiotoxicity, radionuclides are classified into four categories: high toxicity; medium toxicity (sub-group A); medium toxicity (sub-group B); and low toxicity.

7.5.14 Controlled area An area in which individual exposure of personnel to radiation is controlled and which is under the supervision of a person who has knowledge of the appropriate radiation protection regulations and responsibility for applying them.

7.5.15 Dosimetry The measurement or evaluation of the absorbed dose, exposure, dose equivalent or similar radiation quantity.

7.5.16 Intake The quantity of activity entering the body from the external environment.

7.5.17 Ionising radiation Any radiation consisting of directly or indirectly ionising particles or a mixture of both.

Note. In the fields of regulation and radiation protection, visible and ultraviolet light are usually excluded.

7.5.18 Radioactive contamination A radioactive substance in a material or place where it is undesirable.

7.5.8 Equivalent de dose maximale admissible (EDMA) Le plus grand équivalent de dose absorbée, reçu en un temps déterminé permis par un ''Comité de réglementation'' sur la base d'une hypothèse selon laquelle il n'y a pas de probabilité appréciable d'apparition de dommages sommatiques ou génétiques.

Différents niveaux d'équivalent de dose admissible peuvent être fixés pour des groupes différents d'une population.

7.5.9 Dose d'irradiation Dose reçue par des personnes, par une partie ou par l'ensemble de la population du fait d'un rayonnement ionisant d'origine naturelle ou artificielle.

7.5.10 Dose individuelle Exposition, dose absorbée ou équivalent de dose, mesurés à un endroit du corps humain considéré comme représentatif.

7.5.11 Equivalent de dose collective Total des équivalents de dose reçue individuellement par les membres d'une collectivité dans des circonstances déterminées.

Remarque : on préfèrera cette notion à celle de ''dose-homme-rem''.

7.5.12 Retombée radioactive Retombée au sol de substances provenant de l'atmosphère.

7.5.13 Radiotoxicité Toxicité liée à l'absorption de matières radio-actives dans l'organisme humain.

Remarque : selon leur radioactivité relative, on classe les radionucléides en quatre catégories de risque : radiotoxicité très élevée, élevée, moyenne et faible.

7.5.14 Zone contrôlée Zone dans laquelle l'exposition individuelle du personnel aux rayonnements, soumise à autorisation, est contrôlée et qui est supervisée par une personne techniquement compétente en matière de réglements de radioprotection et responsable de leur application. application.

7.5.15 Dosimétrie Détermination de la dose absorbée, de l'exposition, de l'équivalent de dose ou d'autre grandeur de rayonnement similaire.

7.5.16 Apport (incorporation) On appelle apport l'absorption par l'organisme humain ou animal de matières radioactives ou d'autres substances toxiques.

7.5.17 Rayonnement ionisant, rayons ionisants Tout rayonnement composé de particules directement ou indirectement ionisantes ou d'un mélange des deux.

Remarque : comme la quantité d'énergie nécessaire pour provoquer l'ionisation par collision dépend également de la nature du gaz ionisé, il n'est pas possible de définir la quantité d'énergie photonique ou particulaire marquant la limite entre rayonnements non ionisants (lumière visible, par exemple) et rayonnements ionisants.

7.5.18 Contamination radioactive Présence d'une substance radioactive dans un milieu ou au contact d'une matière où elle est indésirable.

7.5.8 Aequivalentdosis, höchstzugelassene Die höchstzugelassene Aequivalentdosis ist die höchste Aequivalentdosis, die eine Person nach Empfehlungen zuständiger Kommissionen oder nach gesetzlichen Regelungen innerhalb einer bestimmten Zeitspanne erhalten darf. Dabei wird angenommen, dass bis zum Erreichen der höchstzugelassenen Aequivalentdosis kein merkliches Risiko eines somatischen oder genetischen Schadens entsheht. Für verschiedene Bevölkerungsgruppen können verschiedene Werte der höchstzugelasenen Aequivalentdosis festgesetzt werden.

7.5.9 Strahlenbelastung Die Strahlenbelastung ist die Belastung von Personen, Bevölkerungsgruppen oder der Gesamtbevölkerung durch ionisierende Strahlung natürlichen oder künstlichen Ursprungs.

7.5.10 Personendosis Die Personendosis ist die Ionendosis, Energiedosis oder Aequivalentdosis an einer für die Strahlenbelastung als repräsentativ geltenden Stelle der Körperoberfläche einer Person.

7.5.11 Gruppen-Aequivalentdosis Summe der Equivalentdosen, die die Angehörigen einer Personengruppe unter bestimmten Umständen erhalten. Anmerkung : Der Begriff soll anstelle von "man-rem-Dosis" verwendet werden.

7.5.12 Niederschlag, radioaktiver Radioaktiver Niederschlag ist Niederschlag radioaktiver Stoffe aus der Atmosphäre auf die Erdoberfläche.

7.5.13 Radiotoxizität Die Radiotoxizität bezeichnet die Gefährlichkeit radioaktiver Stoffe für den Menschen bei ihrer Aufnahme in den menschlichen Körper.

Anmerkung : Entspreched ihrer relativen Radiotoxizität werden Radionuklide in vier Gefahrenklassen eingeteilt : sehr hohe, hohe, mittlere und niedrige Radiotoxizität.

7.5.14 Kontrollbereich Der Kontrollbereich ist ein abgegrenzter Bereich, in dem ein genehmigungspflichtiger Umgang mit radioaktiven Stoffen oder eine genehmigungspflichtige Erzeugung ionisierender Strahlen erlaubt ist und in dem wegen der dort möglichen Strahlenexposition Zustrittund Aufenthaltsbeschränkungen sowie Ueberwachungen nach bestehenden gesetztlichen Vorschriften erforderlich sind.

7.5.15 Dosimetrie Dosimetrie ist die Bestimmung der durch ionisierende Strahlen in Materie erzeugten Energiedosis, Ionendosis oder Aequivalentdosis.

7.5.16 Inkorporation Inkorporation ist die Aufnahme offener radioaktiver oder anderer toxischer Stoffe in einen menschlichen oder tierischen Körper.

7.5.17 Strahlung, ionisierende; ionisierende Strahlen Ionisierende Strahlung ist Strahlung, die unmittelbar (direkt) oder mittelbar (indirekt) durch Stoss zu ionisieren vermag.

Anmerkung : Eine bestimmte Photonen- oder Teilchenenergie als Grenze zwischen nichtionisierender Strahlung (z.B. sichtbarem Licht) und ionisierender Strahlung lässt sich nicht angeben, da die zur Stossionisation benötigte Energie auch von der Art des ionisierten Gases abhängt.

7.5.18 Kontamination, radioaktive Die radioaktive Kontamination ist eine Verunreinigung durch radioaktive Substanzen.

7.5.8 Máxima dosis equivalente admisible Máxima dosis equivalente recibida en un tiempo determinado, permitida por un "Comité de reglamentación" basándose en una hipótesis, según la cual no hay probabilidad apreciable de aparición de daños somáticos o genéticos.

Se pueden fijar diferentes dosis equivalentes máximas admisibles para grupos diferentes de una población.

7.5.9 Dosis de irradiación Dosis debida a la incidencia de una radiación ionizante artificial o artificial y natural en las personas, grupos de la población o la población total.

7.5.10 Dosis individual Exposición, dosis absorbida, o dosis equivalente, medida en un punto del cuerpo humano que se considera representativo.

7.5.11 Dosis equivalente colectiva Total de las dosis equivalentes recibidas individualmente por los miembros de una colectividad en circunstancias determinadas.

Observacion: Se preferirá esta noción a la de "dosis hombre-rem".

7.5.12 Lluvia radiactiva Caída al suelo de sustancias radiactivas procedentes de la atmósfera.

7.5.13 Radiotoxicidad Toxicidad ligada a la absorción de materias reactivas en el organismo humano. *Observación:* Según su radiactividad relativa se clasifican los radionucleidos en cuatro categorias, a causa del riesgo: radiotoxicidad muy alta, alta, media y débil.

7.5.14 Zona controlada Area en la que la exposición individual de una persona a la radiación está controlada y en la que está bajo la supervisión de un experto, quien tiene conocimientos de las reglas de protección apropiadas para la radiación y es responsable de su aplicación.

7.5.15 Dosimetría Medida o evaluación de la dosis absorbida en la exposición, dosis equivalente o de otra magnitud similar de radiación.

7.5.16 Incorporación Absorción. por el organismo humano o animal, de materias-radiactivas o de otras sustancias tóxicas.

7.5.17 Radiación ionizante, rayos, ionizantes Toda radiación compuesta de partículas directa o indirectamente ionizantes o de una mezcla de ambas.

Observación: Como la cantidad de energía necesaria para provocar la ionización por colisión depende igualmente de la naturaleza del gas ionizado, no es posible definir la cantidad de energía fotónica o particular que marca el límite entre las radiaciones no ionizantes (luz visible, por ejemplo) y radiaciones ionizantes.

7.5.18 Contaminación radiactiva Presencia de una sustancia radiactiva en materiales o en ambientes donde su presencia no es deseable.

7

7.5.19 Decontamination Removal or reduction of radioactive contamination, by chemical or physical processes.

7.5.20 Maximum permissible concentration The level of activity concentration of a nuclide present in air, water or foodstuffs which by legal regulation is established as the maximum that would not create undue risk to human health.

7.5.21 Discharge of radioactive materials The controlled emission of radioactive materials into the atmosphere or into waters in the operation of nuclear installations.

7.6   Treatment of Radioactive Wastes

7.6.1 Radioactive waste Unwanted radioactive materials obtained in the processing or handling of radioactive materials, or after their utilisation.

7.6.2 Radioactive waste management The management of radioactive wastes with a view to their controlled ultimate disposal, including for example, concentration, solidification, sealing in containers, storage at an intermediate site.

7.6.3 Concentration processes Processes for reducing the bulk of radioactive wastes, e.g. evaporation, precipitation, incineration.

7.6.4 Intermediate storage site A site where radioactive wastes are stored under controlled conditions prior to their transport to a site for ultimate disposal.

7.6.5 Solidification processes Processes for embodying radioactive wastes in compact solid bodies, e.g. concrete, bitumen or glass.

7.6.6 Ultimate radioactive waste disposal site A site at which radioactive wastes are stored under controlled conditions, such that no further handling is required.

7.5.19 Décontamination Elimination ou réduction d'une contamination radioactive, par des procédés chimiques ou physiques.

7.5.20 Concentration maximale admissible Valeur limite, fixée légalement, de la concentration d'activité d'un nucléide présent dans l'eau ou dans l'air. Cette valeur est un maximum qui peut être sans risque pour la santé de l'homme.

7.5.21 Rejet d'effluents radioactifs Émission contrôlée de substances radioactives dans l'atmosphère ou dans les eaux, résultant du fonctionnement d'installations nucléaires.

7.6   Traitement des déchets Radioactifs

7.6.1 Déchets radioactifs Matières radioactives indésirables obtenues lors du traitement ou de la manipulation de matériels radioactifs.

7.6.2 Gestion des déchets Traitement de déchets radioactifs en vue de les stocker sous contrôle dans une aire de stockage définitive consistant, par exemple, à les concentrer, les solidifier, les placer dans des contenus spéciaux ou à les stocker provisoirement.

7.6.3 Procédés de compactage Procédés permettant de réduire le volume des déchets radioactifs, par exemple par évaporation, précipitation, incinération.

7.6.4 Aire de stockage provisoire Endroit dans lequel les déchets sont conservés sous surveillance avant d'être transportés vers l'aire de stockage définitive.

7.6.5 Procédés de solidification Procédé de fixation ou d'enrobage des déchets radioactifs dans des corps solides compacts (béton, goudron ou verre, par exemple).

7.6.6 Aire de stockage définitive L'aire de stockage définitive est un lieu permettant le stockage-contrôle des déchets radioactifs sous une forme telle qu'il n'est plus nécessaire de les déplacer.

7.5.19 Dekontamination Dekontamination ist die Beseitigung oder Verringerung einer radioaktiven Kontamination mittels chemischer oder physikalischer Verfahren.

7.5.20 Aktivitätskonzentration, maximal zulässige (MZK) Die maximal zulässige Aktivitätskonzentration ist ein in gesetzlichen Vorschriften für die menschliche Gesundheit angegebener Grenzwert für die Aktivitätskonzentration eines Nuklids in Luft oder Wasser.

7.5.21 Ableitung radioaktiver Stoffe Die Ableitung radioaktiver Stoffe ist die kontrollierte Abgabe radioaktiver Stoffe in die Luft und in Gewässer beim Betrieb kerntechnischer Anlagen.

7.6 Behandlung radioaktiver Abfälle

7.6.1 Radioaktiver Abfall Radioaktiver Abfall ist nichtverwendbarer, radioaktiver Stoff, der bei der Aufarbeitung oder nach der Benutzung von radioaktiven Stoffen anfällt.

7.6.2 Abfallbehandlung Abfallbehandlung ist die Behandlung radioaktiver Abfälle mit dem Ziel der kontrollierten Unterbringung in einem Endlager z.B. Konzentrieren, Verfestigen, Einschliessen in Behälter, Zwischenlagerung.

7.6.3 Konzentrierungsverfahren Konzentrierungsverfahren sind Verfahren zur Verringerung des Volumens radioaktiver Abfälle, z.B. Eindampfen, Ausfällen, Veraschen.

7.6.4 Zwischenlager Ein Zwischenlager ist ein Lager, in dem radioaktiver Abfall unter kontrollierten Bedingungen vor seinem Transport in das Endlager aufbewahrt wird.

7.6.5 Verfestigungsverfahren Verfestigungsverfahren sind Verfahren zum Ueberführen oder Einbetten radioaktiver Abfälle in kompakte feste Körper, z.B. in Beton, Bitumen oder Glasmassen.

7.6.6 Endlager Das Endlager ist ein Ort, an dem radioaktiver Abfall unter kontrollierten Bedingungen so gelagert werden kann, dass eine weitere Verbringung nicht erforderlich ist.

7.5.19 Descontaminación Eliminación o reducción de contaminación radiactiva mediante procedimientos químicos o físicos.

7.5.20 Máxima concentración admisible Valor límite, fijado legalmente, de la concentración de actividad de un nucleido presente en el agua o en el aire. Es un valor máximo que no supone riesgo para la salud humana.

7.5.21 Descarga de matriales radiactivos Emisión controlada de materiales radiactivos en el ambiente o en las aguas en el funcionamiento de instalaciones nucleares.

7.6. Tratamiento de los Residuos Radiactivos

7.6.1 Residuos radiactivos Materiales radiactivos indeseables obtenidos en el tratamiento o la manipulación de los materiales radiactivos.

7.6.2 Gestión de residuos Tratamiento de residuos radiactivos a fin de almacenarlos definitivamente bajo control en un cementerio radiactivo, y consistente, por ejemplo, en concentrarlos solidificarlos, encerrarlos en recipientes especiales o almacenarlos provisionalmente.

7.6.3 Metodo de concentración Procedimientos para reducir el volumen de los residuos radiactivos, por ejemplo por evaporación, precipitación o incineración.

7.6.4 Zona de almacenamiento provisional Lugar en el que los residuos radiactivos se conservan bajo vigilancia, antes de ser transportados a la zona de almacenamiento definitivo.

7.6.5 Método de solidificación Procedimiento para incorporar los residuos radiactivos en cuerpos sólidos y compactos, por ejemplo en hormigón asfalto y vidrio.

7.6.6 Zona de almacenamiento definitivo Lugar que permite el almacenamiento controlado de los desechos radiactivos bajo una forma tal que no es necesario volver a desplazarlos.

7

Section 8

Impact of Energy Industries on the Environment
Influence des industries de l'énergie sur l'environnement
Umweltbeeinflussung durch Energiewirtschaft
Efectos de las Industrias Energeticas en el Ambiente

8

Impact of Energy Industries on the Environment

8.1 General Terms

8.1.1 **Environmental pollution** Any detrimental alteration to the environment caused by man and affecting human, animal and plant life.

8.1.2 **Polluter pays principle** The principle that those causing environmental harm by producing or utilising energy shall bear the cost of its remedy, i.e. such cost shall become a component in the cost of the product.

8.1.3 **Site criteria** Those criteria that must be applied when deciding on the suitability of a location for a plant. From the point of view of the environment, such criteria would apply to the effect of the construction and operation of the plant on the health and safety risk to the local population, on levels of air, water and soil pollution and on local amenities in general. Factors to be considered would include: existing levels of pollution, cooling capacity of waters, population density, regional economic structure, topography, geology, nature of regional development, wind direction, earthquake risks.

8.2 Material Pollution of the Environment

8.2.1 **Air pollution** Any detrimental alteration of the ambient air caused by man.

8.2.2 **Emission** The release of substances or energy (e.g. noise, vibration, radiation, heat) into the environment from a source.

8.2.3 **Ground level concentration (of pollutants); immission (SA)** The incidence of substances or energy (e.g. noise, vibration, radiation, heat) in a specified place, whereby environmental conditions for man, plants and animals become detrimentally altered.

8.2.4 **Pollution source; pollution emitter** Plant or equipment that releases substances or energy (e.g. noise, vibration, radiation, heat) into the environment.

8.2.5 **Emission point; point source** A point at which substances or energy are released into the environment.

8.2.6 **Emission standard** A level of emission that under the law may not be exceeded.

8.2.7 **Ambient air (water etc.) quality standard; immission standard** (SA) A level of ambient air (water, etc.) quality that under the law must be maintained.

8.2.8 **TLV (threshold limit value) at place of work; occupational TLV; MAC (maximum allowable concentration) at place of work; occupational MAC** The maximum allowable concentration of air polluting substances that, in continued and generally daily 8-hour impact on the human organism, is not damaging to health.

8.2.9 **TLV (threshold limit value) in the free environment; MAC (maximum allowable concentration) in the free environment** The maximum concentration of air polluting substances in the free environment whose impact when of specified duration and frequency is not objectionable to man, fauna and flora.

8.2.10 **Instantaneous concentration** Regional ambient pollutant concentration at a specified point in time.

Influence des industries de l'énergie sur l'environnement

8.1 Termes généraux

8.1.1 **Dégradation de l'environnement** Détérioration, provoquée par l'homme, des conditions de vie pour les personnes, les animaux et les plantes.

8.1.2 **Principe pollueur-payeur** Principe selon lequel le responsable de dommages ou de toute atteinte à l'environnement doit répondre financièrement des mesures réparatoires.

8.1.3 **Critères d'implantation** Ensemble des facteurs à considérer lors du choix du lieu d'implantation d'une installation énergétique, par exemple : niveau de pollution, capacité de refrigération, densité de population, structure économique, topographie, géologie, aménagement du territoire, direction des vents, risques sismiques.

8.2 Pollution des Milieux

8.2.1 **Pollution atmosphérique** Toute modification défavorable causée par l'homme à l'air ambiant.

8.2.2 **Emission** Libération de substances ou d'énergie (bruit, vibration, radio-activité, chaleur) dans un milieu.

8.2.3 **Nuisance** Présence de substances ou d'énergie (bruit, vibration, radio-activité, chaleur) dans un milieu donné, modifiant les conditions ambiantes pour les hommes, les animaux ou les plantes.

8.2.4 **Source de pollution** Installation ou dispositif qui libère des substances ou des énergies (bruit, vibration, radio-activité, chaleur) dans un milieu.

8.2.5 **Lieu d'émission** Lieu où des substances ou de l'énergie sont libérées dans un milieu.

8.2.6 **Limite d'émission** Niveau d'émission qui ne doit pas être dépassé, selon la réglementation.

8.2.7 **Limite de nuisance** Niveau de nuisance qui ne doit pas être dépassé, selon la réglementation.

8.2.8 **Concentration maximale admissible dans les ambiances profesionnelles** Teneur limite d'un polluant atmosphérique qui, de façon générale, n'affecte pas la santé d'une personne qui y est soumise quotidiennement pendant huit heures.

8.2.9 **Concentration maximale admissible dans l'air ambiant** Teneur limite d'un polluant atmosphérique qui est considérée comme sans inconvénient pour les hommes, les animaux et les plantes lorsqu'il agit pendant une durée et à une fréquence déterminée.

8.2.10 **Niveau instantané de nuisance** Concentration locale d'un polluant atmosphérique à un instant déterminé.

Umweltbeeinflussung durch Energiewirtschaft

8.1 Allgemeine Begriffe

8.1.1 Umweltbelastung Eine von Menschen verursachte Verschlechterung der Lebensbedingungen für Menschen, Tiere und Pflanzen.

8.1.2 Verursacherprinzip Grundsatz, dass der Verursacher für die Massnahmen zur Beseitigung von Umweltschäden oder Umweltbeeinträchtingungen verantwortlich ist.

8.1.3 Standortkriterien Gesamtheit der Faktoren, die bei der Wahl des Standortes einer Energieanlage zu berücksichtigen sind, z.B. Vorbelastung, Kühlleistung, Bevölkerungsdichte, Wirtschaftsstruktur, Topographie, Geologie, Raumplanung, Windrichtung, Erdbebenhäufigkeit.

8.2 Stoffliche Beeinflussung der Umwelt

8.2.1 Luftverunreinigung Vom Menschen verursachte nachteilige Veränderung der Umgebungsluft.

8.2.2 Emission Abgabe von Stoffen und Energie (z.B. Schall, Erschütterung, Strahlung, Wärme) aus einer Quelle an die Umwelt.

8.2.3 Immission Auftreten von Stoffen und Energien (z.B. Schall, Erschütterung, Strahlung, Wärme) an einem bestimmten Ort, wodurch die Umweltverhältnisse für Mensch, Tier und Pflanze verändert werden.

8.2.4 Emittent Anlage oder Einrichtung, die Stoffe und Energie (z.B. Schall, Erschütterung, Strahlung, Wärme) an die Umwelt abgibt.

8.2.5 Emissionsquelle Stelle des Austritts von Stoffen oder Energie in die Umwelt.

8.2.6 Emissionsgrenzwert Durch Vorschriften festgelegter Wert für eine Emission, der nicht überschritten werden darf.

8.2.7 Immissionsgrenzwert Durch Vorschriften festgelegter Wert für eine Immission, der nicht überschritten werden darf.

8.2.8 MAK-Wert Höchstzulässige Konzentration luftverunreinigender Stoffe, die bei fortgesetzer und in der Regel täglich acht-stündiger Einwirkung auf den menschlichen Organismus nicht gesundheitsschädigend ist.

8.2.9 MIK-Wert Höchste Konzentration luftverunreinigender Stoffe in der freien Atmosphäre, die für Mensch, Tier und Pflanze bei Einwirkung von bestimmter Dauer und Häufigkeit als unbedenklich gilt.

8.2.10 Grundbelastung Die zu einem bestimmten Zeitpunkt gemessene regionale Immission.

Efectos de las Industrias Energeticas en el Ambiente

8.1 Conceptos Generales

8.1.1 Degradación del ambiente Deterioro, provocado por el hombre de las condiciones de vida que afecta a las personas, los animales y las plantas.

8.1.2 Principio "el que contamina paga" Principio en virtud del cual los causantes de perjuicios o de cualquier atentado al medio ambiente deben responder financieramente de la medidas para su corrección.

8.1.3 Criterios de emplazamiento Conjunto de factores a tener en cuenta a la hora de decidir el emplazamiento de una instalación energética, por ejemplo: nivel de contaminación, capacidad de refrigeración, densidad de población, estructura económica, topografía, geología, ordenación del territorio, dirección de los vientos, riesgos sísmicos.

8.2 Contaminación de los Ambientes

8.2.1 Contaminación atmosférica Toda modificación de carácter desfavorable causada por el hombre al aire ambiental.

8.2.2 Emisión Descarga de sustancia o de energía (por ejemplo, ruido, vibración, radiactividad, calor) en el medio ambiente.

8.2.3 Inmisión Incidencia de sustancias y energía (ruido, vibración, radioactividad, calor) en lugar determinado, de suerte que quedan alteradas las condiciones del medio ambiente para el hombre, plantas y animales.

8.2.4 Fuente de contaminación Instalación o equipo que descarga sustancias o energías (ruido, vibración, radioactividad, calor) en el medio ambiente.

8.2.5 Punto de emisión Punto en el que se descargan sustancias o energía al medio ambiente.

8.2.6 Límite de emisión Nivel de emisión que, de acuerdo con las leyes, no debe ser rebasado.

8.2.7 Límite de inmisión Nivel de inmisión del ambiente que, de acuerdo con las leyes, no debe sobrepasarse.

8.2.8 Concentración máxima admisible en los ambientes de trabajo Concentración máxima de un contaminante atmosférico que, de manera general no pone en peligro la salud de una persona que se encuentre sometida a sus efectos durante 8 horas diarias.

8.2.9 Concentración máxima admisible en el aire ambiente Concentración máxima de un contaminante atmosférico que se estima no perjudica al hombre, fauna y flora, si se trata de una duración y frecuencia determinadas.

8.2.10 Concentración instantánea de contaminación Concentración local de un contaminante atmosférico en un momento determinado.

8

8.2.11 **Air pollutants** Substances that detrimentally alter the natural composition of the air.

8.2.12 **Exhaust; waste gases** Gases vented to atmosphere by stationary plant and appliances and by internal combustion engine driven vehicles.

8.2.13 **Gaseous combustion products; flue gases** Gases that arise in combustion processes.

8.2.14 **Fumes** A concentration of solid and liquid particles, and visible gases in a carrier gas.

8.2.15 **Steam-laden emissions; mists; plumes** Air, supersaturated with water vapour and often containing solid, liquid or gaseous contaminants, that is vented from industrial processes.

8.2.16 **Dust** Solid particulate matter in disperse phase in a carrier gas.

8.2.17 **Dust content** The mass or number of solid particles dispersed in unit volume of carrier gas.

8.2.18 **Soot** Fine-grained particulate amorphous carbon occurring as the result of incomplete combustion.

8.2.19 **Smog** A concentration of air pollutants occurring under particular meteorological conditions, generally of photo-chemical origin.

8.2.20 **Aerosols** Suspended particles colloidally dispersed in a gaseous medium.

8.2.21 **Plume-rise** The difference between the height of the emission point and the height to which the emissions rise due to thermally or kinetically induced buoyancy.

8.2.22 **Water pollution** Any detrimental alteration of surface waters, underground waters or the marine environment caused by man.

8.2.23 **Water pollutants** Solid, liquid and gaseous substances that detrimentally alter the natural condition of waters.

8.2.24 **Aqueous effluent** Water discharged after use by households, trade or industry, also polluted rainwater from an inhabited area.

8.2.25 **Initial level of water pollution** The material and thermal pollution of water prior to the use under consideration.

8.2.26 **Water quality** The classification of surface waters categorising their degree of pollution.

8.2.27 **Turbid water** Colloid disperse systems in water.

8.2.28 **Land pollution; soil pollution** Any detrimental alteration to land or soil caused by man.

8.2.29 **Land pollutants; soil pollutants** Solid, liquid and gaseous substances that detrimentally alter the natural condition of the soil or land.

8.2.30 **Contaminant fall-out; fall-out; deposition** The quantity of polluting material precipitated from the atmosphere per unit area per unit time.

8.2.31 **Wash-out; atmospheric scrubbing** The cleansing of the atmosphere by natural precipitation (rain or snow) entraining airborn contaminants to the surface of the earth.

8.2.32 **Usable by-products; usable waste products** Solid or liquid, storable, residual materials of varying consistencies to which a value can be attached.

8.2.33 **Combustion residue; ash** Inert or unburned matter remaining after a process of combustion.

8.2.11 **Polluants atmosphériques** Substances solides, liquides ou gazeuses qui modifient de façon défavorable la composition de l'air ambiant.

8.2.12 **Effluents gazeux** Gaz qui sont rejetés à l'atmosphère par les équipements d'installation et, par les vehicules de transport munis de moteur à explosion et à combustion.

8.2.13 **Gaz de combustion** Gaz résultant d'opérations de combustion.

8.2.14 **Fumée** Rejet de produits gazeux, rendus visibles par les particules solides et liquides qu'ils entraînent.

8.2.15 **Panaches** Masse d'air contenant de la vapeur d'eau sursaturée et souvent aussi d'autres produits solides, liquides ou gazeux, rejetée par une installation industrielle.

8.2.16 **Poussières** Particules solides qui peuvent être entraînées par un gaz dans lequel elles sont dispersées.

8.2.17 **Teneur en poussières** Masse ou nombre de particules solides dispersées dans l'unité de volume du gaz porteur.

8.2.18 **Suie** Fines particules de carbone amorphe résultant d'une combustion incomplète.

8.2.19 **Smog** Apparition d'une forte concentration de polluants atmosphériques dans des conditions météorologiques déterminées.

8.2.20 **Aérosol** Particules en suspension colloïdale dans un milieu gazeux.

8.2.21 **Ascension des panaches** Différence entre le niveau du point d'émission et le niveau auquel s'élèvent les polluants par suite du gradient thermique ou de leur énergie cinétique.

8.2.22 **Pollution des eaux** Toute modification défavorable causée par l'homme aux eaux de surface, aux eaux souterraines ou au milieu marin.

8.2.23 **Polluants des eaux** Substances solides, liquides ou gazeuses qui modifient de façon défavorable l'état d'une eau.

8.2.24 **Eaux usées** Eaux rejetées après emploi domestique, commercial ou industriel et eaux de pluies polluées provenant d'une zone habitée.

8.2.25 **Niveau initial de pollution** Pollution matérielle ou calorifique d'une eau avant l'utilisation envisagée.

8.2.26 **Niveaux de qualité des eaux** Désignation des catégories d'eaux de surface d'après leur degré de pollution.

8.2.27 **Boues** Matières en dispersion colloïdale dans l'eau.

8.2.28 **Pollution des sols** Toute modification défavorable causée par l'homme à des terrains.

8.2.29 **Polluants des sols** Substances solides, liquides ou gazeuses qui modifient de façon défavorable l'état naturel d'un sol.

8.2.30 **Précipitation de polluants** Quantité de polluants provenant de l'atmosphère et se déposant sur une surface-unité de sol pendant une durée-unité.

8.2.31 **Condensation atmospherique** Épuration de l'air par les précipitations atmosphériques.

8.2.32 **Déchets utilisables** Matières résiduaires solides ou liquides de consistances diverses qui peuvent être stockées et valorisées.

8.2.33 **Imbrûlés** Matières non combustibles . ou non brûlées résultant d'une opération de combustion.

8.2.11 Luftuerunreinigenda Stoffe Stoffe, die die natürliche Zusammensetzung der Luft nachteilig verändern.

8.2.12 Abgase In die atmosphäre geleitete abgase aus stationären und beweglichen quellen.

8.2.13 Verbrennungsgas Gas, das bei Verbrennungsvorgängen entsteht.

8.2.14 Schwaden Verdichtung von festen und flüssigen Teilchen und sichtbaren Gasen in einem Trägergas.

8.2.15 Brüden Mit Wasserdampf übersättigte - oft auch feste, flüssige oder gasförmige Verunreinigungen enthaltende - bei technischen Prozessen entweichende Luft.

8.2.16 Staub In einem Trägergas dispergierte Feststoffe.

8.2.17 Staubgehalt Masse oder Teilchenzahl des in der Volumeneinheit des Trägergases vorhandenen Staubes.

8.2.18 Russ Feinverteilter amorpher Kohlenstoff als Folge unvollständiger Verbrennung.

8.2.19 Smog Entstehung und Verdichtung luftverunreinigender Stoffe bei bestimmten meteorologischen Bedingungen.

8.2.20 Aerosole Kolloidal verteilte Schwebeteilchen in einem gasförmigen Dispersionsmittel.

8.2.21 Ueberhöhung Differenz zwischen der Höhe der Emissionsquelle und der Aufstiegshöhe der Emissionen, wobei das Aufsteigen durch den thermischen und kinetischen Auftrieb verursacht wird.

8.2.22 Wasserverunreinigung Vom Menschen verursachte nachteilige Veränderung von Oberflächen-, Grund- oder maritimen Gewässern.

8.2.23 Wasserverunreinigende Stoffe Feste, flüssige und gasförmige Stoffe, die die natürliche Beschaffenheit eines Gewässer verändern.

8.2.24 Abwasser Durch häuslichen, gewerblichen oder industriellen Gebrauch verändertes Wasser sowie verschmutztes Niederschlagswasser aus dem Bereich menschlicher Ansiedlungen.

8.2.25 Vorbelastung Stoffliche und thermische Belastung des Wassers vor der jeweiligen Nutzung.

8.2.26 Gewässergüteklassen Begriff zur Klassifizierung der Oberflächengewässer zur Beurteilung des Verschmutzungsgrades.

8.2.27 Trübe Kolloid-disperses System im Wasser.

8.2.28 Bodenverunreinigung Vom Menschen verursachte nachteilige Veränderung des Erdbodens.

8.2.29 Bodenverunreinigende Stoffe Feste, flüssige und gasförmige Stoffe, die die natürliche Beschaffenheit des Bodens verändern.

8.2.30 Fremdstoffniederschlag Die je Flächeneinheit in der Zeiteinheit aus der Atmosphäre niedergeschlagene Stoffmenge.

8.2.31 Auswaschvorgang, atmosphärisch Auswaschen der Luft durch Niederschläge.

8.2.32 Verwertbare Abfallprodukte (Abprodukte) Feste oder flüssige lagerungsfähige Rückstände verschiedener Konsistenz, die einer Verwertung zugeführt werden können.

8.2.33 Verbrennungsrückstände Nicht brennbare und unverbrannte Stoffe, die nach einem Verbrennungsprozess zurückbleiben.

8.2.11 Contaminantes de la atmósfera Sustancias sólidas, líquidas o gaseosas que alteran de forma perjudicial la composición del aire ambiente.

8.2.12 Effluentes gaseosos Gases, descargados a la atmósfera por los equipos de las instalaciones y por los vehículos de transporte provistos de motor de explosion y de combustión.

8.2.13 Gases de combustión Gases resultantes de los procesos de combustión.

8.2.14 Humo Descarga de productos gaseosos, que resultan visibles gracias a las partículas sólidas y líquidas que arrastran.

8.2.15 Penachos (emisiones de vapor, neblinas) Masa de aire, sobresaturado con vapor de agua y conteniendo a menudo contaminantes sólidos, líquidos o gaseosos, vertido a la atmósfera por una instalación industrial.

8.2.16 Polvo Partículas sólidas en fase dispersa en un gas portador.

8.2.17 Contenido en polvo Masa o número de partículas dispersadas en unidad de volumen de gas portador.

8.2.18 Hollía Partículas finas de carbón amorfo, resultantes de una combustión incompleta.

8.2.19 Smog Concentración intensa de contaminantes de la atmósfera que se presenta bajo condiciones meteorológicas determinadas.

8.2.20 Aerosoles Partículas en suspensión coloidal, en un medio gaseoso.

8.2.21 Elevación del penacho Diferencia entre la altura del punto de emisión y la altura a la cual se elevan los contaminantes debido al gradiente térmico o a su energía cinética.

8.2.22 Contaminación del agua Toda modificación perjudicial causada por el hombre a las aguas superficiales, o subterráneas o al medio marino.

8.2.23 Contaminantes del agua Sustancias sólidas, liquidas y gaseosas que alteran perjudicialmente la condición natural de las aguas.

8.2.24 Aguas residuales Aguas vertidas después de su uso doméstico, comercial o industrial, incluidas las aguas pluviales contaminadas, procedentes de una zona habitada.

8.2.25 Nivel inicial de contaminación de las aguas Contaminación material o calorífica del agua con anterioridad al uso a que vaya a ser destinada.

8.2.26 Nivel de calidad de las aguas Clasificación de las aguas superficiales según su grado de contaminación.

8.2.27 Fangos, lodos, aguas turbias Comprende los sistemas dispersos de coloides en el agua.

8.2.28 Contaminación del suelo Toda modificación perjudicial causada sobre los terrenos por el hombre.

8.2.29 Contaminantes del suelo Sustancias sólidas, líquidas o gaseosas que alteran perjudicialmente la condición natural del suelo.

8.2.30 Precipitacion de contaminantes Cantidad de sustancias contaminantes precipitadas desde la atmósfera por unidad de superficie y de tiempo.

8.2.31 Condensación atmosférica Purificación del aire gracias a las precipitaciones atmosféricas.

8.2.32 Desechos utilizables Sustancias residuales sólidas o liquidas, almacenables, de consistencias variables a las que se puede atribuir un valor.

8.2.33 Residuos de combustión (inquemados) Materias combustibles o no quemadas, resultantes de un proceso de combustión.

8

8.3 Thermal Pollution of the Environment

8.3.1 Waste heat Heat energy that has not been utilised in an industrial thermal process and is released to the surrounding air, soil or waters.

8.3.2 Thermal load The waste heat absorbed by waters, soil or the atmosphere.

8.3.3 Temperature rise; incremental heating The difference between the outlet and inlet temperatures of the cooling medium in a cooling system.

8.3.4 Heat load plan A plan of existing and future thermal loading of waters, soil or the atmosphere so as to maintain biological equilibrium.

8.4 Noise Pollution of the Environment

8.4.1 Sound pressure level The logarithm of the ratio of effective sound pressure to reference sound pressure (in the region of the threshold of audibility). In English-speaking countries the unit of sound pressure level is the decibel which is defined as 20 times the logarithm of the ratio of the sound pressure in question to the reference sound pressure.

8.4.2 Sound pressure spectrum The sound pressure level plotted as a function of frequency.

8.4.3 Sound source; noise source Any solid, liquid or gaseous system or medium able to vibrate in the range of audible frequencies.

8.4.4 The metered and weighted sound pressure level at a specific place over a specific period of time.

Note. There is no exact English equivalent for this term. The terms "noise dose" or "noise immission level" may convey the meaning of this term in certain contexts, but where precision is required the term used should be defined, as certain commonly employed terms do not yet have established definitions.

8.5 Radioactive Pollution of the Environment

8.5.1 The incidence of man-made or natural ionizing radiation on persons, groups of the population or the whole population. The word "exposure" may convey the meaning of this term when used in a loose sense, but care should be taken not to confuse this meaning with the precise meaning of the term exposure as defined in 7.5.5.

8.5.2 Ionizing radiation See definition 7.5.17.

8.5.3 Maximum permissible concentration See definition 7.5.20.

8.5.4 Dose See definition 7.5.2

8.5.5 Dose equivalent See definition 7.5.7.

8.5.6 Quality factor A factor depending on the linear energy transfer in water of primary or secondary charged particles, by which absorbed dose is multiplied to obtain, according to practice in the field of radiation protection, an evaluation on a common scale, for all ionizing radiations, of the irradiation incurred by exposed persons.

8.5.7 Maximum permissible dose equivalent (MPDE) See definition 7.5.8.

8.3 Pollution Thermique

8.3.1 Chaleurs perdues Énergie non utilisée au cours d'une opération thermique industrielle et rejetée dans le milieu sous forme de chaleur.

8.3.2 Apports thermiques Quantité de chaleurs perdues absorbées par les eaux, les sols ou l'atmosphère.

8.3.3 Echauffement Différence entre la température de sortie et la température d'entrée d'un fluide dans un dispositif de réfrigération.

8.3.4 Objectif de pollution thermique Politique des apports thermiques, actuels et futurs, aux eaux, aux sols ou à l'atmosphère, destinée à sauvegarder les équilibres biologiques.

8.4 Nuisances Acoustiques

8.4.1 Niveau sonore (exprimé en décibels) Logarithme du rapport de la pression acoustique effective à la pression acoustique de référence (seuil d'audition).

8.4.2 Spectre acoustique Représentation graphique du niveau sonore en fonction de la fréquence du bruit.

8.4.3 Source de bruit Tout objet ou milieu solide, liquide ou gazeux susceptible de vibrer dans le domaine des fréquences audibles.

8.4.4 Indice d'évaluation du bruit Niveau sonore mesuré et évalué indiviuellement pendant une durée déterminée à un emplacement donné.

8.5 Pollution radioactive

8.5.1 Irradiation Rayonnements ionisants d'origine naturelle ou artificielle reçus par un individu, un groupe ou par la population mondiale.

8.5.2 Rayonnements ionisants Voir définition 7.5.17.

8.5.3 Concentration maximale admissible Voir définition 7.5.20.

8.5.4 Dose de rayonnement Voir définition 7.5.2.

8.5.5 Equivalent de dose Voir définition 7.5.7.

8.5.6 Facteur de qualité (protection contre les rayonnements) Dans le calcul des équivalents de dose, produit de facteurs correctifs par lequel doit être multipliée la dose de rayonnements pour évaluer les risques radioactifs des différents types de rayonnements ionisants en fonction des conditions d'exposition.

8.5.7 Equivalent de dose maximale admissible Voir définition 7.5.8.

8.3 Thermische Beeinflussung der Umwelt

8.3.1 Abwärme Die in einem wärmetechnischen Prozess nicht genutzte Energie, die als Wärme an die Umgebung abgeführt wird.

8.3.2 Wärmebelastung Aufnahme von Abwärme durch Gewässer, Boden oder die Atmosphäre.

8.3.3 Aufwärmspanne Die Differenz zwischen Aus- und Eintrittstemperatur des Kühlmittels in einem Kühlsystem.

8.3.4 Wärmelastplan Plan der bestehenden und zukünftigen Wärmebelastung von Gewässern, Boden oder Atmosphäre, zur Aufrechterhaltung des biologischen Gleichgewichts.

8.4 Akustische Beeinflussung der Umwelt

8.4.1 Schallpegel (Schalldruckpegel) Logarithmus des Verhältnisses des effektiven Schalldruckes zum Bezugsschalldruck (Hörschwelle).

8.4.2 Schallspektrum (Schalldruckspektrum) Darstellung des Schalldruckpegels als Funktion der Frequenz.

8.4.3 Schallquelle (Schallerzeuger) Alle im Bereich der Schallfrequenzen schwingenden festen, flüssigen oder gasförmigen Gebilde oder Medien.

8.4.4 Lärmbelastung Der in einer bestimmten Zeitdauer an einem bestimmten Ort gemessene und bewertete Schallpegel.

8.5 Radioaktive Beeinflussung der Umwelt

8.5.1 Strahlenbelastung Belastung von Personen, Bevölkerungsgruppen oder der Gesamtbevölkerung durch ionisierende Strahlung natürlichen oder künstlichen Ursprungs.

8.5.2 Strahlung, ionisierende Siehe Definition 7.5.17

8.5.3 Aktivitätskonzentration, maximal zulässige (MZK) Siehe Definition 7.5.20

8.5.4 Strahlendosis Siehe Definition 7.5.2

8.5.5 Aequivalentdosis Siehe Definition 7.5.7

8.5.6 Bewertungsfaktor (Strahlenschutz) Produkt von festzulegenden Faktoren, mit denen bei der Berechnung der Aequivalentdosis die Energiedosis multipliziert wird, um dem jeweiligen Strahlenrisiko für die verschiedenen Arten ionisierender Strahlung und den jeweiligen Strahlungsbedingungen Rechnung zu tragen.

8.5.7 Aequivalentdosis, höchstzugelassene Siehe Definition 7.5.8

8.3 Contaminación térmica

8.3.1 Calor residual Energía calorífica que no ha sido utilizada en un proceso industrial térmico y es descargada a la atmósfera, suelo o aguas circundantes, en forma de calor.

8.3.2 Cargas térmicas Cantidad de calores residuales absorbidos por las aguas, suelos o atmósfera.

8.3.3 Elevación de temperatura Es la diferencia entre las temperaturas a la entrada y salida de un medio refrigerante en un sistema de refrigeración.

8.3.4 Programa de carga térmica Programa de aportaciones térmicas existentes y futuras, a las aguas, a los suelos o a la atmósfera a fin de salvaguardar el equilibrio biológico.

8.4 Contaminación acústica

8.4.1 Nivel sonoro (expresado en decibeles) Logaritmo de la relación entre la presión acústica efectiva y la presión acústica de referencia (umbral de audición).

8.4.2 Espectro acústico Representación gráfica del nivel sonoro en función de la frecuencia del ruido.

8.4.3 Emisor de ruido; foco de ruido Todo objeto o medio sólido, líquido o gaseoso susceptible de vibrar en el dominio de las frecuencias audibles.

8.4.4 Indice de evaluación del ruido Nivel sonoro, medido y evaluado, en un lugar determinado, con una duración determinada.

8.5 Contaminación radiactiva

8.5.1 Irradiación Radiaciones ionizantes de origen natural o artificial recibidas por un individuo, por un grupo o por la población mundial.

8.5.2 Radiaciones ionizantes Vease definición 7.5.17.

8.5.3 Concentración máxima admisible de productos radiactivos Véase definición 7.5.20

8.5.4 Dosis de radiación Vease definición 7.5.2

8.5.5 Dosis equivalente Vease definición 7.5.7

8.5.6 Factor de calidad (protección contra las radiaciones) En el cálculo de las dosis equivalentes, es el producto de los factores correctores por el que debe ser multiplicada la dosis de radiación a fin de evaluar los riesgos radiactivos de los diversos tipos de radiaciones ionizantes en función de las condiciones de exposición.

8.5.7 Máxima dosis equivalente admisible Véase definición 7.5.8

8.5.8 Population dose A measure of the total exposure of the whole body or a specified organ of a population of people in terms of dose equivalent. The population dose is given in rems.
Note. For a fuller definition see: Publication No. 22 of the International Commission on Radiological Protection.

8.5.9 Group/sub-population collective dose A component of the population dose (see 8.5.8 above) related to a given sub-population, which, for some purposes, may be the population of a country or region. The group/sub-population collective dose is measured in rems.

8.5.10 Intake See definition 7.5.16.

8.5.11 Radioactive fall-out The deposition upon the surface of the earth of radioactive substances from the explosion of a nuclear device or from their accidental release.

8.5.12 Radioactive waste See definition 7.6.1.

8.6 Topographical Impact on the Environment

8.6.1 Acquisition of land The acquisition of any right or interest in land (including land covered by buildings, trees or water) whether by agreement or by the exercise of compulsory powers and whether by the Crown, a public or local authority, nationalised industry or any other person, corporate or incorporate. A *concession* is a special case of the acquisition of land, whereby a lease or licence is granted by agreement to operators to prospect for and work mineral deposits.

8.6.2 Land disturbance; land degradation The destructive working of, or dumping on, unspoilt land, with particular reference to surface mining activities.

8.6.3 Rehabilitation (of land and waters) The conversion of surface areas (lands and waters), after their commercial utilisation for energy purposes is completed, to commercially useful lands, forests and waters or to other useful purposes. This comprises reclamation and subsequent recultivation.

8.6.4 Reclamation (of land and waters) Measures taken to render surface areas (lands and waters), formerly utilised for energy purposes, fit for recultivation.

8.6.5 Recultivation Measures taken to ensure the permanent and productive utilisation of reclaimed areas.

8.6.6 Controlled dumping The properly regulated deposition of waste products.

8.5.8 Dose totale pour une population Équivalents de dose totaux reçus par l'ensemble d'une population.

8.5.9 Equivalents de dose pour un groupe Somme des équivalents de dose reçus par les individus d'un groupe dans des conditions déterminées.

8.5.10 Apport Voir définition 7.5.16.

8.5.11 Retombées radioactives Dépôt de matières radioactives provenant de l'atmosphère sur la surface de la terre.

8.5.12 Déchets radioactifs Voir définition 7.6.1.

8.6 Influence de l'Environnement sur l'Aménagement du Territoire

8.6.1 Concession Autorisation règlementaire d'utiliser des terrains (sol et eau) objet d'une exploitation agricole, forestière ou autre, à des fins énergétiques.

8.6.2 Dévastation Enlèvement et/ou recouvrement d'un sol naturel (sol sans perturbation artificielle).

8.6.3 Remise en valeur (d'un terrain) Reconversion des terrains (sol et eau) après leur utilisation à des fins énergétiques en vue d'une exploitation agricole ou forestière ou de l'aménagement des eaux ou toute autre utilisation. Cette reconversion comprend notamment la remise en état, puis en culture des sols.

8.6.4 Remise en état (d'un terrain), redéfrichement Mesures prises en vue de rendre aptes à la culture des terrains (sol et eau) après leur utilisation à des fins énergétiques.

8.6.5 Remise en culture (d'un terrain) Ensemble de mesures prises en vue d'assurer une production agricole durable sur des terrains reconvertis.

8.6.6 Dépôt de déchets contrôlé Dépôt ordonné de produits de déchets.

8.5.8 Bevölkerungsgesamtdosis Auf die Gesamtbevölkerung bezogen Gruppenäquivalentdosis.

8.5.9 Gruppen-Aequivalentdosis Summe der Aequivalentdosen, die die Angehörigen einer Personengruppe unter bestimmten Umständen erhalten.

8.5.10 Inkorporation Siehe Definition 7.5.16

8.5.11 Niederschlag, radioaktiver Niederschlag radioaktiver Stoffe aus der Atmosphäre auf die Erdoberfläche.

8.5.12 Abfall, radioaktiver Siehe Definition 7.6.1

8.6 Territoriale Beeinflussung der Umwelt

8.6.1 Flächenentzug Gesetzlich geregelte Inanspruchnahme land-, forst-, wasserwirtschaftlicher oder sonstiger Nutzflächen für energiewirtschaftliche Zwecke.

8.6.2 Devastierung Abbau oder Ueberkippen gewachsenen Bodens.

8.6.3 Wiedernutzbarmachung Ueberführung von Flächen (Boden und Gewässern) nach Aufgabe ihrer energiewirtschaftlichen Nutzung in land-, forst- und wasserwirtschaftliche sowie sonstige Nutzung. Sie umfasst Wiederurbarmachung und nachfolgendes Rekultivieren.

8.6.4 Wiederurbarmachung Massnahmen, um ehemals energiewirtschaftlich genutzte Flächen (Boden und Gewässer) für land-, forst- und wasserwirtschaftliche sowie sonstige Zwecke rekultivierbar zu machen.

8.6.5 Rekultivierung Folgemassnahmen, um wiederurbargemachte Flächen einer nachhaltigen und ertragssicheren Nutzung zuzuführen.

8.6.6 Deponie Geordnete Ablagerung von Abfallprodukten (Abprodukten).

8.5.8 Dosis total de una población Dosis equivalentes totales recibidas por el conjunto de una población.

8.5.9 Dosis equivalentes de un grupo Suma de las dosis equivalentes recibidas por los individuos de un grupo en determinadas condiciones.

8.5.10 Incorporación Véase definición 7.5.16.

8.5.11 Lluvia radiactiva Precipitación, sobre la superficie terrestre, de materias radiactivas procedentes de la atmósfera.

8.5.12 Residuos radiactivos Vease definicion 7.6.1.

8.6 Influencia del ambiente sobre la Ordenacion del Territorio

8.6.1 Concesión Autorización reglamentaria para la utilización, con fines energéticos, de los terrenos (suelo y agua) objeto de una explotación agrícola, forestal o de alguna otra clase.

8.5.2 Devastación Desmonte o terraplenado de un suelo natural (suelo sin perturbación artíficial).

8.6.3 Reconstitución (de un terreno) Reconversión de los terrenos (suelo y agua) con vistas a una explotación agrícola o forestal, al aprovechamiento de las aguas, o cualquier otra finalidad útil, después de haber sido utilizado con fines energéticos. Esta reconversión incluye principalmente las operaciones de regulación de los terrenos y posterior transformación en tierras cultivables.

8.6.4 Reposición (de un terreno) Medidas a tomar para hacer apropiados al cultivo los terrenos (suelo y agua) previamente utilizados con finnes energéticos.

8.6.5 Recultivación (de un terreno) Conjunto de medidas a tomar con el fin de asegurar una explotación agrícola permanente sobre los terrenos reconstituidos.

8.6.6 Vertido controlado Depósito autorizado de residuos.

8

Alphabetical Index

The first digit of the number following a term indicates the section within which the term falls, i.e.: 1, general; 2, electrical; 3, hydro; 4, solid fuels; 5, liquid fuels; 6, gaseous fuels; 7, nuclear; 8, environmental.

Index Alphabetique

Le premier chiffre de l'indice de référence, après chaque terme, indique la section dans laquelle se trouve ce dernier, par exemple : 1, concepts généraux; 2, industrie électrique; 3, energie hydroélectrique; 4, combustibles solides; 5, combustibles liquides; 6, industrie gazière; 7, industrie nucléaire; 8, environnement.

Index

Die erste Ziffer des Indexes nach jedem Begriff bezieht sich auf das entsprechende Begriffsfeld z.B. : 1,Allgemeine Begriffe; 2, Elektrizitätswirtschaft; 3, Wasserkraftwirtschaft; 4, Gewinnung und Verarbeitung fester Brennstoffe; 5, Gewinnung und Verarbeitung flüssiger Brennstoffe; 6, Begriffe der Gaswirtschaft; 7, Kernenergiewirtschaft; 8, Umweltbeeinflussung durch Energiewirtschaft.

Indice Alfabetico

La primera cifra del indice de referencia, a continuación de cada término, indica la sección en que dicho término se encuentra; es - decir: 1, conceptos generales; 2, industria eléctrica; hidroeléctrica; 4, combustible sólidos, 5, combustibles liquidos; 6, industria del gas; 7, industria nuclear; 8, medio ambiente.

133